Embedded Systems Design Based on Formal Models of Computation

Ivan Radojevic • Zoran Salcic

Embedded Systems Design Based on Formal Models of Computation

Ivan Radojevic
Defence Technology Agency
New Zealand Defence Force
Auckland
New Zealand
ivan_radojevic@hotmail.com

Zoran Salcic
Computer Systems Engineering
University of Auckland
Auckland
New Zealand
z.salcic@auckland.ac.nz

ISBN 978-94-017-8415-3 ISBN 978-94-007-1594-3 (eBook)
DOI 10.1007/978-94-007-1594-3
Springer Dordrecht Heidelberg London New York

Softcover re-print of the Hardcover 1st edition 2011

Printed on acid-free paper

Springer is part of Springer Science+Business Media (www.springer.com)

Preface

One of the key problems in modern embedded systems design is the productivity gap. While the performance of computing architectures has been rapidly increasing in the past few decades, design tools have not kept pace. As a result, it is becoming increasingly difficult for embedded systems designers to handle complex applications. Delays in product delivery and even project cancellations are quite common. An obvious solution is to raise the abstraction level of design tools and at the same time enable automatic synthesis from high level specifications. A complete embedded application needs to be specified in a single system level language rather than using programming languages and hardware description languages to create an early non-optimal hardware/software partition. From a single system specification written in a formal language, it is possible to automate design space exploration, hardware/software partitioning, verification and synthesis, which gives an enormous boost to design productivity. However, all these design activities can be automated by design tools only if the system-level specification is constructed according to a formal model of computation, which sets the rules for communication among concurrent processes comprising the system. While several models of computation have been successfully used in certain classes of applications, their applicability in complex embedded system design is quite limited. In particular, there is a lack of suitable models for heterogeneous embedded systems that contain both control-driven and data-driven behaviours.

This book offers a new design methodology for design of heterogeneous embedded systems. At the heart of the methodology lies a model of computation called DFCharts. A complete design flow is covered, from a system specification in a formal language to an implementation on a multiprocessor architecture. Throughout the book examples are provided to illustrate main concepts. The reader is not required to have a deep understanding of models of computation. Only basic familiarity is assumed. For this reason, following the introductory Chaps. 1 and 2 describes a number of widely used models of computation and system level languages. Chaps. 3–8 then present the DFCharts based design methodology. The conclusion with future research directions follow in Chap. 9.

New Zealand

Ivan Radojevic
Zoran Salcic

Contents

List of Figures

List of Tables

Chapter 1
Introduction

1.1 Embedded Systems Design

An embedded computing system represents a set of processing elements embedded inside a larger system and usually communicates with the physical world. Although embedded systems are already widespread, the number of applications is expanding both in traditional areas such as communications, consumer electronics, aerospace, automotive, and in new ones such as biomedical.

Embedded systems differ in a number of ways from general purpose computing systems. An embedded system must be able to work at a rate imposed by the environment, which is not a requirement for a general purpose computing system. Concurrency and timing issues must be dealt with in embedded systems design. Furthermore, a range of constraints has to be faced. Some of the usual ones are performance, cost and power. While a general purpose computing system involves only software, an embedded system is often a mixture of software and hardware parts. Software runs on traditional microprocessors, digital signal processors (DSP) and various application specific processors (ASP). Hardware parts are implemented with application specific circuits (ASIC) or field programmable gate arrays (FPGA).

Traditionally, embedded system design starts with an informal specification. A decision is immediately made on how the functionality is going to be split between software and hardware. From this point, software and hardware parts are designed and verified separately. Software is written with programming languages like C/C++ and Java. Hardware description languages (HDL) like VHDL and Verilog are used for the design of the hardware part of the system. The major difficulty comes when interfaces between software and hardware parts are created. At this stage many small implementation details must be handled because there is no standard way in which software and hardware communicate. Many problems are discovered, which sometimes necessitate a complete redesign. Moreover, it is difficult to verify that the final implementation satisfies the original specification.

I. Radojevic and Z. Salcic, *Embedded Systems Design Based on Formal Models of Computation*, DOI 10.1007/978-94-007-1594-3_1,

The traditional design methodology, which is still used in most embedded designs, may be satisfactory for smaller and medium sized systems. However, it is becoming increasingly difficult to use for large systems. The functionality and complexity of embedded systems keeps growing. Technology advances make it possible to integrate an increasing number of components on a single chip. Moore's law, which states that the number of transistors on a chip doubles every 18 months, still holds. On the other hand, design methods are improving much slower. This results in a problem, known as productivity gap [1].

Design at higher levels of abstraction is a promising solution to overcome the productivity gap [2]. It should start with a specification created with a formal language. A variety of languages have been proposed for that purpose [3]. A specification should capture only the behaviour of a system without any reference to implementation. Various names have been used to refer to this level of abstraction such as system, behavioural, algorithmic or functional level. As much as possible of the system verification should take place at the system-level. The system-level verification is much faster than the verification at lower levels of abstraction since many low level implementation details are not visible. Simulation is still the main verification method, but formal verification [4] is becoming important especially for safety critical embedded systems. After the verification has been completed, the final implementation consisting of software and hardware parts should ideally be synthesized automatically.

The behaviour of an embedded system is usually captured as a set of concurrent, communicating processes. The rules that are used for computation inside processes and communication between processes are determined by a *model of computation* [5, 6]. The computation inside each process is sequential. A finite state machine (FSM) [7] can be used to describe it, or more expressive models could be used that may have the full power of Turing machines. It is much more difficult to decide on the concurrency model that dictates the communication between processes. Currently, most embedded software design is done with threads which communicate using mutual exclusions locks, semaphores and interrupts. These communication mechanisms are difficult to use and often result in bugs that are hard to detect [8]. Sometimes a design may work for years before bugs show up and it suddenly crashes.

The alternatives to threads are various concurrency models with formal semantics. They employ different scheduling policies and communication mechanisms resulting in different orderings of events, which trigger or are associated with system operations. The most frequently used ones are discrete event [9], asynchronous dataflow [10], synchronous reactive [11], Petri nets [12] and process algebras such as communicating sequential processes (CSP) [13] and calculus of communicating systems (CCS) [14]. All these models impose some restrictions on communication but in return provide useful properties. The designer would have most freedom if communication between processes were unrestricted. However, tools would not be able to analyse such specifications and automated synthesis would not be possible. Instead, a specification would have to be refined manually to a lower level of abstraction. Because of manual refinement, it would be necessary to verify that the behaviour has been preserved.

In general there is a trade-off between analyzability and expressiveness in a model of computation. A successful model needs to strike a balance. It needs to provide a framework for analysis but at the same time it needs to be expressive enough.

An important feature of embedded systems behaviour is heterogeneity. Two major types of embedded systems behaviour can be identified: control-dominated and data-dominated. Control-dominated systems have to quickly react to events that arrive from the external environment at irregular and random time instances. A lift controller and vehicle automation are examples of control-dominated systems. Reaction time is less important for data-dominated systems which operate on samples that arrive at regular intervals. However, data-dominated systems perform computations that are much more intensive than those in control-dominated systems. Most of digital signal processing algorithms, such as FFT, FIR and IIR filters, are data-dominated. Most embedded systems contain both control-dominated and data-dominated behaviours. For example, a mobile phone has to perform data-dominated operations such as speech processing, coding, modulation but it also has to take care of control-dominated operations such as network protocols or reacting to user commands.

With the integration of analogue parts together with digital parts on a single chip, heterogeneity of embedded systems will become even more pronounced. Models that aim to address these mixed signal systems need to be able to support continuous time. In this book, we are concerned with purely digital systems. Thus, when we refer to heterogeneous embedded systems, we mean systems that represent a mixture of data-dominated and control dominated parts.

Well established models of computation have specific advantages but are not able to successfully handle entire heterogeneous embedded systems. Asynchronous dataflow models [10] consist of processes that communicate through FIFO channels with blocking reads. Using blocking reads ensures that outputs do not depend on the scheduling order and speeds of processes. Dataflow models have been successfully applied in signal processing and other transformational systems. However, they lack reactivity because of blocking reads. In the synchronous/reactive model [11], all processes read inputs and produce outputs simultaneously when a clock tick occurs. Blocking reads are not necessary for determinism. This feature is evident in synchronous language Esterel [15], which is deterministic, but has plenty of reactive statements. Synchronous/reactive model has also had success in the signal processing area with languages Lustre [16] and Signal [17], which have dataflow flavour. The requirement that all events in a system are synchronous can have a big implementation price. This especially applies in the design of large systems on chip. It may mean that all processing elements in a system must wait for each other at some point even though they are performing unrelated tasks. In the context of a pure hardware implementation, distributing a single clock can be a problem.

Modelling heterogeneous embedded systems is still an open research area with plenty of room for advancement. This is witnessed by the absence of mature, well-developed tools for heterogeneous systems. Esterel studio [18] is a commercial design environment from Esterel technologies, which uses Esterel language for creating specifications. It is very convenient for design of control-dominated systems but lacks features to support data-dominated behaviour. Another tool from Esterel technologies called SCADE is based around synchronous language Lustre, but it also allows insertion of FSMs to support control-dominated behaviours. The combination of Lustre and FSMs is still completely synchronous. Hence it could have implementation difficulties when applied in the design of large embedded systems. In order to produce efficient implementations, Polis [19] and its commercial successor VCC use a globally synchronous locally asynchronous (GALS) model called Codesign finite state machines (CFSM). However, their target is just control-dominated behaviour. Ptolemy [20] is a graphical environment that supports a variety of models of computation. It is still primarily used for simulation. Automatic synthesis has been demonstrated for synchronous dataflow (SDF) domain, but not for heterogeneous specifications. Simulink [21], while it does have some code generation capabilities, is also primarily a simulation environment. Recently, a heterogeneous model called SysteMOC [122, 123] has emerged. It combines dataflow and finite state machines in a relatively straightforward manner, which allows for efficient design space exploration and synthesis.

In this book, after reviewing most widely used models of computation and languages for embedded systems design, we present an approach for designing heterogeneous embedded systems based on our model of computation called DFCharts. A complete design flow from the specification to the implementation on a multiprocessor architecture is described. The methodology is demonstrated with a practical heterogeneous embedded system applied in power systems monitoring. In addition, we suggest how DFCharts based modelling can be used to improve design with two popular system level languages, SystemC and Esterel.

While DFCharts can be used for specification of distributed systems, it is not highly suitable for this task since it does not have any special features that support distributed processes. For this reason, an extension of DFCharts towards distributed systems called DDFCharts (Distributed DFCharts) has been designed. The semantics of DDFCharts will be described alongside DFCharts in this book. However, the design flow from DDFCharts specifications has not yet been completely developed and it remains an important future research direction.

1.2 DFCharts

DFCharts combines hierarchical concurrent finite state machines (HCFSM) [22] with synchronous dataflow graphs (SDFG) [23]. Three communication mechanisms are employed in DFCharts: synchronous broadcast used between FSMs, FIFO channels used inside SDFGs, and rendezvous channels used between FSMs and SDFGs. The semantics of synchronous broadcast is as in Argos [24], a Statechart [22] variant with purely synchronous communication. An SDFG can be placed anywhere in the FSM hierarchy. Thus, it is possible to specify complex control logic that leads to activation and termination of SDFGs. When expressed with its graphical syntax, the DFCharts model looks as shown in Fig. 1.1.

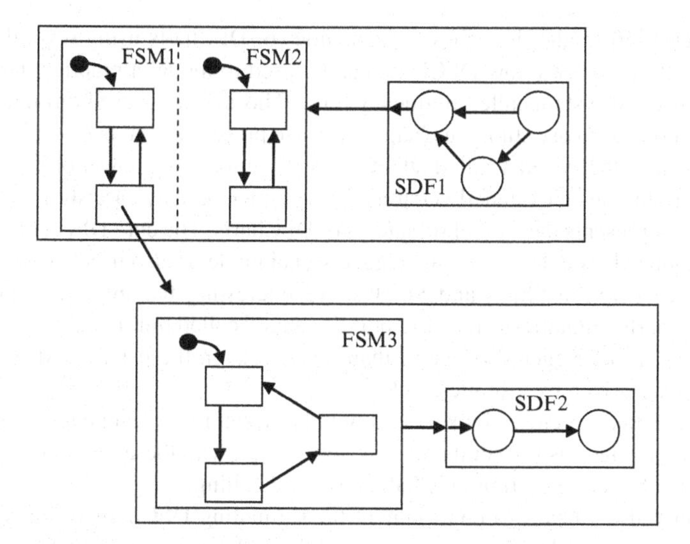

Fig. 1.1 Graphical syntax of DFCharts

We have described the semantics of communication between FSMs and SDFGs in detail within tagged-signal model (TSM) framework [25]. Another type of semantics we use is based on automata. It is similar to the semantics of Argos. It represents the operation of an SDFG as an FSM. In this way a complete DFCharts model can be flattened to produce a single FSM. This in turn allows the model behaviour to be analysed globally. The automata based semantics can handle any dataflow model that has a clearly defined iteration and executes in bounded memory. For example, cyclo-static dataflow (CSDF) [26] can be easily incorporated in DFCharts, but Kahn process network (KPN) [27] would require an extension in the current semantics. We focus on SDF, which is suitable for a large range of applications.

All FSMs in a DFCharts model are driven by a single clock. On the other hand, each SDFG operates at its own speed. Structurally, a DFCharts model resembles an Esterel program with asynchronous tasks where asynchronous tasks are comparable to SDFGs and possibly other dataflow models placed in DFCharts. However, asynchronous tasks in Esterel are essentially part of the external environment, outside the Esterel semantics as we discussed in [28]. They are very vaguely defined. In DFCharts semantics, SDFGs are fully integrated.

1.3 Book Organization

Chapter 2 provides a brief survey of models of computation that are used in embedded systems design. It also covers system level languages, which can be related to models of computation. Some of these languages are based on a single model of computation while others are capable of describing multiple ones.

Chapter 3 informally describes the semantics of DFCharts using several simple examples. It also shows how DFCharts can be used to model a practical heterogeneous embedded system called frequency relay. This system is used throughout the rest of the book as a case study. In Chap. 3, we also present an extension of DFCharts towards distributed systems called DDFCharts. The modelling power of DDFCharts is demonstrated on an extended version of the frequency relay case study.

Chapter 4 presents the formal semantics of DFCharts. The core DFCharts semantics is automata based. However, the tagged signal model is also used for describing the interface between FSMs and SDFGs. An interesting feature of the automata based DFCharts semantics is its basic building block called multiclock FSM, where transitions can be triggered by more than one clock. In the usual FSM model all transitions are driven by a single clock.

Chapter 5 takes a closer look at two popular system level languages, SystemC and Esterel. It suggests how embedded system design with these two languages can be improved by incorporating DFCharts based modelling.

Chapter 6 describes Java environment for simulating DFCharts designs. While the Java packages underpinning execution and synchronization of FSMs are completely new, the simulation of SDFGs is done by invoking Ptolemy. There are very few interfacing problems in this solution since Ptolemy software has also been designed in Java.

Chapter 7 provides an overview of the current state of the art multiprocessor architectures and points to their deficiencies in implementing heterogeneous embedded systems. It then defines a new type of architecture called HETRA which has special features for supporting heterogonous embedded systems. The subset of HETRA called HiDRA has a major role in the DFCharts based design flow.

Chapter 8 describes in detail the DFCharts based design flow for heterogeneous embedded systems. It starts from a high level specification, which is refined into lower level processes that are mapped on HiDRA processing units. A major strength of the DFCharts based design flow is the ability to support a trade-off between implementation efficiency and verification effort. This is quite useful in the design of large heterogeneous embedded systems.

Finally, Chap. 9 presents a conclusion and some open future directions in research.

Chapter 2
Models of Computation and Languages

This chapter looks at several important models of computation and languages for embedded systems design. We do not attempt to draw sharp distinctions between models and languages. Thus, the topics in this section are not divided strictly to either models of computation or languages. A model of computation is commonly used for defining the semantics of a language. However, when a model of computation is expressed with a simple syntax, it can also be called a language. For example, synchronous dataflow (SDF) is usually thought of as a model of computation. But as soon as it is expressed with a simple graphical syntax consisting of a network of blocks connected with arrows, it is not incorrect to call it a language.

2.1 Finite State Machine

An FSM [7] can be defined as a tuple of six elements:

- Q is a finite set of states
- Σ is a set of input symbols
- Δ is a set of output symbols
- δ is a transition function mapping $Q \times \Sigma$ to $Q \times \Delta$
- q0 is the initial state

An FSM reacts to inputs by entering the next state and producing outputs as defined by the transition function δ. The state transition diagram is a common way of representing an FSM. A simple state transition diagram is shown in Fig. 2.1.

In this case $Q = \{A,B\}$; $\Sigma = \{c,d,e\}$; $\Delta = \{x,y\}$; $\delta(A,c) = (x,B)$, $\delta(B,d) = (A,y)$ and $\delta(B,e) = (B,y)$, $q0 = A$.

States are denoted by circles. Transitions are denoted by arcs. Every transition is associated with a guard and an action. This is labelled as *guard/action* on the corresponding arc. *Guard* denotes the enabling input event $i \in \Sigma$ that causes a transition

I. Radojevic and Z. Salcic, *Embedded Systems Design Based on Formal Models of Computation*, DOI 10.1007/978-94-007-1594-3_2,

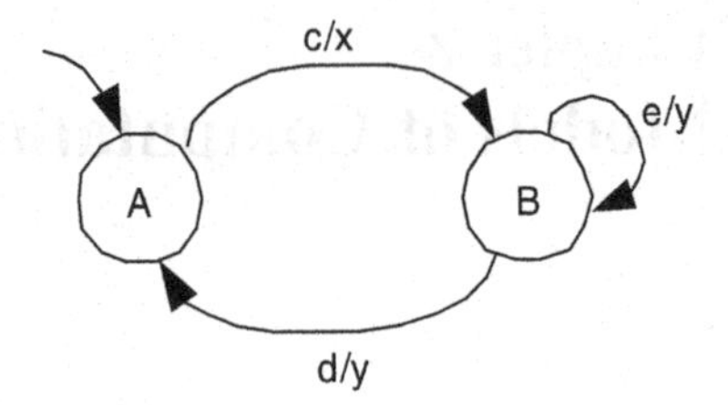

Fig. 2.1 Simple state transition diagram

Current state	A	A	B	B	....
Input symbol	d	c	e	d	....
Next state	A	B	B	A	....
Output symbol	-	x	y	y	....

Fig. 2.2 Possible trace for FSM in Fig 3.1

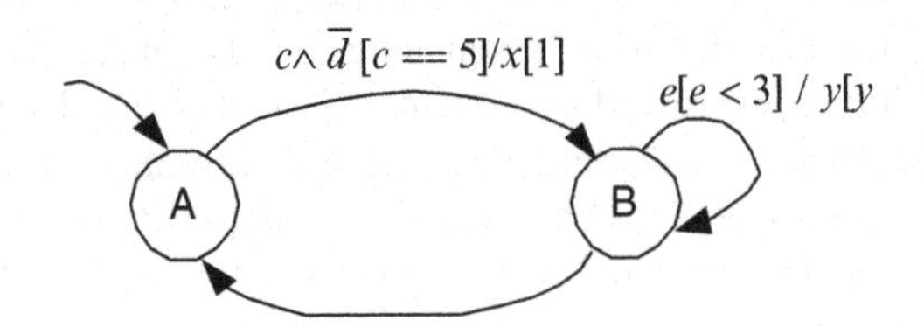

Fig. 2.3 FSM with valued events

from one state to another. *Action* denotes the output event o ∈ Δ that is produced as result of the transition.

If a non-enabling input event occurs in a certain state, the implicit self-transition is assumed. A sequence of reactions, sometimes called *trace*, shows a sequence of states and a sequence of output symbols caused by a sequence of input symbols. A possible trace for the FSM in Fig. 2.1 is shown in Fig. 2.2

The FSM model is closely related to the finite automata (FA) model. An FA is designed to recognize whether a sequence of input symbols belongs or does not belong to a certain language. Thus it accepts or rejects a sequence of input symbols. An FSM also produces a sequence of output symbols in addition to changing states.

Input and output events can be *pure* or they can carry a *value*. If an event is pure, it can only be *present* or *absent*. A valued event, besides being present or absent, also has a value when it is present.

Valued events increase the expressiveness of an FSM. A Boolean expression representing the guard of a transition can contain values of events. Values of output events can be expressed in terms of arithmetic operations. An example of a state transition diagram with valued events is given in Fig. 2.3.

In the above example, " = = " is used for input events to test equality, whereas "=" is used for output events as an assignment operator. A Boolean expression in a

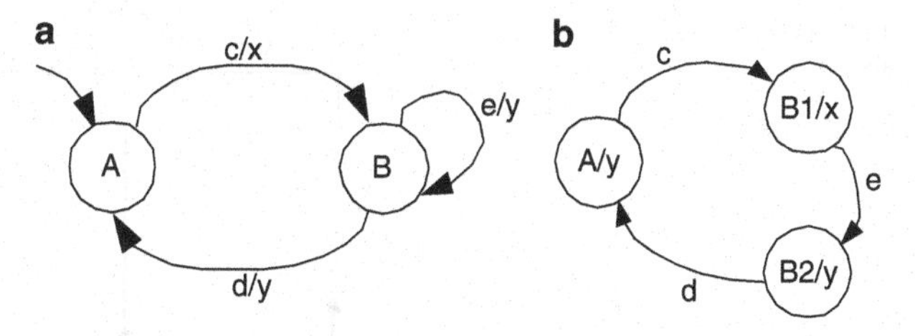

Fig. 2.4 Mealy and Moore machines

transition guard may require that some events be absent for the transition to take place. In Fig. 2.3, $c \wedge \overline{d}[c = 5]$ means that the transition occurs if c is present and has the value of 5 and d is absent.

Generally, two types of FSMs are commonly used. A *Mealy machine* is an FSM where outputs are associated with the transition. A *Moore machine* is an FSM where outputs are associated with the present state of the FSM. The FSMs presented so far are Mealy machines since their outputs are associated with transitions.

A Moore machine may have more states than the equivalent Mealy machine. Figure 2.4 shows the Mealy machine (part (a)) from Fig. 2.1 with the equivalent Moore machine (part (b)).

FSMs can be specified formally in a clear way. Their strong formal properties make them attractive for safety critical applications. It is easier to avoid undesirable states with FSMs then with if-else, goto and other statements found in programming languages.

A single FSM could hardly be used to cover the entire control behaviour of a larger system because it would have an impractically large number of states. The usefulness of FSMs was largely increased when Harel introduced Statecharts (described in Sect. 2.7). In Statecharts, a single FSM state can be refined to another FSM. Thus, a hierarchical description of system behaviour is possible. The other major innovation in Statecharts is the possibility of having two or more states that are active at the same time. This helps in describing concurrent behaviours. A simple flat FSM has no means to describe hierarchy and concurrency, the two features that are commonly found in embedded systems.

2.2 Kahn Process Networks

Synchronous dataflow (SDF), which is used in DFCharts, belongs to the group of dataflow models in which processes communicate through first-in-first-out (FIFO) channels using blocking reads. In order to understand the properties of SDF, we need to explore the most general model in the dataflow group, called Kahn Process Networks (KPN) [27].

KPN processes communicate through FIFO channels using blocking reads and non-blocking writes. At any point during the execution of a Kahn process network, a process

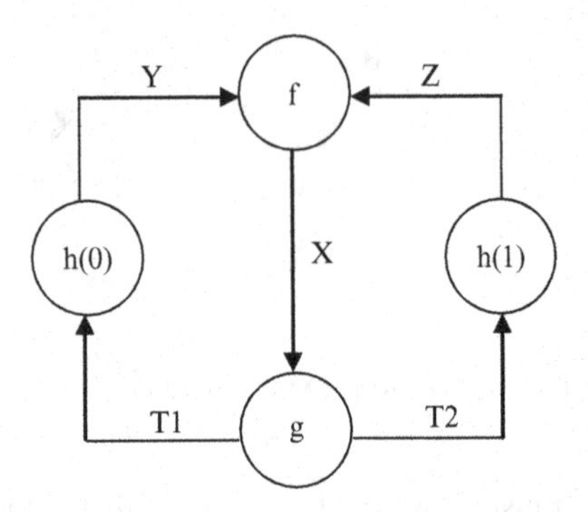

Fig. 2.5 A Kahn process network example

can either be waiting for an input or doing computations. Since blocking reads are used, a process cannot test a channel for the presence of data. If a process attempts to read from an empty channel it will become blocked until data arrives on the channel.

Figure 2.5 shows a KPN example that was used in [27]. The two instances of process *h* copy data from input to output, but initially they output 0 and 1. Without these initial tokens the network would be deadlocked from the start. Process *g* copies data from input channel X to output channels T1 and T2, alternately. Process *f* copies data alternately from input channels Y and Z to output channel X.

In the denotational semantics of Kahn process networks, processes are mathematically defined as functions that map potentially infinite input streams to output streams. A stream is a sequence of data elements $X=[x_1, x_2, x_3, x_4 \ldots]$, where indices are used to specify temporal features of the sequence. The empty sequence is marked by the symbol $\perp$. A relation on streams called *prefix ordering* [10] is useful for analyzing the mathematical properties of Kahn process networks. For example the sequence X=[0] is a prefix of the sequence Y=[0,1] which is in turn a prefix of Z=[0,1,2]. The relation "X is a prefix of Y or equal to Y" is written as $X \subseteq Y$.

A *chain* is an ordered set in which any two elements are comparable. Alternate names for a chain are *linearly ordered set* and *totally ordered set* [29]. In the context of Kahn process networks, elements of a chain are sequences and they are compared with the prefix ordering relation $\subseteq$. Any increasing chain $\vec{X} = (X_1, X_2 \ldots)$ with $X_1 \subseteq X_2 \subseteq \ldots$ has a least upper bound $\Pi\vec{X}$ (symbol Π is used to denote a least upper bound). A least upper bound is a sequence whose length tends towards infinity:

$$\lim_{i \to \infty} X_i = \Pi\vec{X}$$

The set of all sequences is a complete partial order (c.p.o.) with the relation $\subseteq$, since any increasing chain of sequences in this set has a least upper bound.

In Kahn process networks, a process f maps input streams to output streams. A process f is continuous if and only if for any increasing chain $\vec{X} = (X_1, X_2 \ldots)$:

$$f(\Pi\vec{X}) = \Pi f(\vec{X})$$

If a process is continuous, it is also monotonic (the opposite is not necessarily true). Monotonicity means that:

$$X \subseteq Y \Rightarrow f(X) \subseteq f(Y)$$

A process network can be described with a set of equations, with one equation for each process. For example, the process network in Fig. 2.5 can be represented by the following set of equations:

$$T_1 = g_1(X), T_2 = g_2(X), X = f(Y,Z), \; Y = h(T_1,0), \; Z = h(T_2,1)$$

The system of equations above can be reduced to a single equation. For instance the equation for X is:

$$X = f\left(h\left(g_1(X),0\right), h\left(g_2(X),1\right)\right)$$

If all processes are continuous the set of equations has a unique least fixpoint solution. The solution represents the histories of tokens that appeared on the communication channels. For example, the solution for X is an infinite sequence of alternating 0's and 1's – X=f(Y,Z)=[0,1,0,1 …]. The proof by induction can be found in [27].

The blocking read semantics of Kahn process networks ensures that processes are continuous. Therefore a set of equations describing a Kahn process network will have a unique least fixed point solution. This leads to a very useful property of KPN. Any execution order of processes will yield the same solution i.e. the same histories of tokens on the communication channels.

While the execution order cannot influence the histories of tokens, it can greatly impact memory requirements (buffer sizes). Since writes are non-blocking, there are no restrictions on buffer sizes. There are two major methods for scheduling Kahn process networks, data-driven scheduling and demand driven scheduling [30]. In data driven scheduling, the semantics of the Kahn process networks is satisfied in a simple way – a process is unblocked as soon as data is available. Data driven scheduling can lead to unbounded accumulation of tokens on communication channels.

An alternative strategy is to use demand driven scheduling of processes, where a process is activated only when the tokens it produces are needed by another process. Kahn and MacQueen describe a demand driven scheduling method in [31]. A process that needs tokens is marked as *hungry* and that causes the producer of those tokens to be activated. That, in turn, can cause another activation. All the scheduling is done by a single process.

Regardless of the type of scheduling employed, decisions in KPN have to be made at run time. Thus, context switching becomes inevitable if multiple processes run on a single processor. Valuable time has to be spent on saving the state of the current thread before the control can be transferred to another thread.

2.3 Synchronous Dataflow

Synchronous dataflow (SDF) [23, 32] imposes limitations on KPN in order to make static scheduling possible. An SDF network is composed of *actors* that are connected by FIFO channels. When an actor fires, it consumes tokens from input channels and produces tokens on output channels. Firings of an SDF actor create a process. The firing rule of an actor specifies how many tokens are consumed on each input. In SDF, the constant number of tokens is consumed on each input in every firing i.e. the firing rule remains the same. It should also be emphasised that an SDF actor has to output a constant number of tokens on each output in every firing. Due to constant consumption and production rates of tokens it is possible to make very efficient static schedules.

Figure 2.6 shows an SDF network that consists of three actors. Consumption and production rates are labelled on each channel. For example, from the direction of the ch1, it can be seen that its production rate is RA1 and its consumption rate is RC1.

There are two steps in constructing a static schedule for an SDF graph. The first step is to determine how many times each actor should fire during an iteration. An iteration is a series of actor firings that return the channels to their original state. The number of tokens in a channel is the same before and after an iteration. The first step is accomplished by solving the set of balance equations [32]. Balance equations state that production and consumption of tokens must be equal on all channels. The balance equations for the SDF graph from Fig. 2.6 are shown below.

$$FA \times RA1 = FC \times RC1$$

$$FA \times RA2 = FB \times RB1$$

$$FB \times RB2 = FC \times RC2$$

FA, FB, FC are integers showing how many times actors A, B and C fire in a single iteration. They form a *firing* or *repetition vector*. The least positive integer solution is taken. For example if RA1 = 2, RA2 = 2, RB1 = 3, RB2 = 3, RC1 = 6 and RC2 = 6 then FA = 3, FB = 2 and FC = 1.

If the only solution to the set of equations is zeros, the SDF graph is said to be inconsistent [33]. This means that production and consumption of tokens cannot be balanced on all channels. As a result, the execution of an inconsistent SDF graph

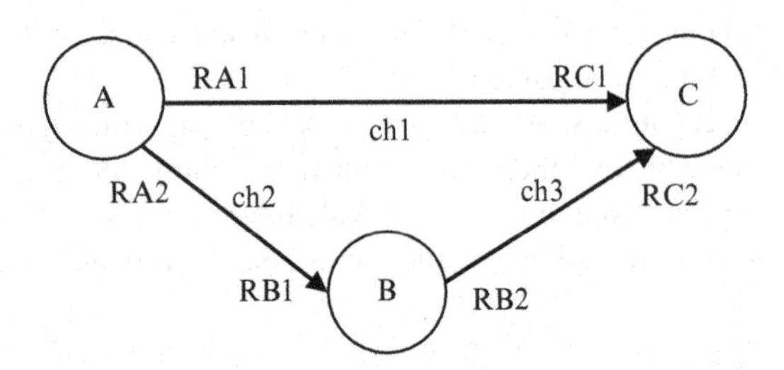

Fig. 2.6 An SDF graph

results in unbounded accumulation of tokens on channels. The graph from Fig. 2.6 would become inconsistent if RC1 were equal to 5, for example.

The second step is to analyse data dependencies between SDF actors in order to determine the order of firings. Multiple valid execution orders can emerge from the analysis due to different interleaving of actor firings. For example, the execution orders AAABBC and AABABC can both be used for the SDF graph from Fig. 2.6.

An SDF graph can contain a cycle that is formed by two or more channels. Initial tokens must be placed on a channel in the cycle so that deadlock does not occur. This was relevant for the KPN example from Fig. 2.5 where the two h processes produced initial values.

SDF is suitable for a wide range of digital signal processing (DSP) systems with constant data rates. It has efficient static scheduling and always executes in bounded memory. These properties are very useful in embedded systems design. For this reason, SDF graphs have been adopted as a part of DFCharts. For systems with variable rates, KPN or dynamic dataflow models can be used. There are many dataflow models whose expressiveness falls between SDF and KPN, such as boolean dataflow (BDF) [34], cyclostatic dataflow (CSDF) [26], parameterized synchronous dataflow (PSDF) [35], multidimensional synchronous dataflow [36], synchronous piggyback dataflow [37] among others.

A large amount of research has been done on synchronous dataflow resulting in numerous techniques and algorithms for memory optimization [38–44], simulation [45, 46], software synthesis [47, 48], hardware synthesis [49–51], and HW/SW codesign [52–54].

2.4 Synchronous/Reactive Model

The synchronous reactive (SR) model of computation [11] is the underlying model for the group of synchronous languages which includes Esterel [15], Argos [24], Lustre [16] and Signal [17]. A brief description of all four languages can be found in [55]. In the SR model of computation, time is divided into discrete instants. In each instant (tick), inputs are read and outputs are computed instantaneously. This is the central assumption in the *synchrony hypothesis* of the SR model. Instantaneous computation and communication makes outputs synchronous to inputs. The status of each signal has to be defined in each tick. It can be either present (true) or absent (false).

This model is similar to synchronous digital circuits that are driven by clocks. As a result, SR models can be efficiently synthesised into hardware. Software synthesis is also possible. In that case, the time between two successive instants is usually not constant.

The assumption of instantaneous computation facilitates hierarchical specification of systems. When a process is broken down into several other processes, they will all have instantaneous computation.

Zero delay communication represents a challenge for compilers of synchronous languages. An SR compiler has to be able to deal with causality loops that arise as a result of zero delays. When resolving the status of each signal in a tick three general outcomes are possible:

- There is a single solution. The signal is either present or absent.
- There is no solution. The model does not make sense.
- Both the presence and absence of the signal satisfy the model. Thus, the system is non-deterministic.

The first outcome is the desired one. The last two outcomes should be rejected by the compiler and an error should be reported to the user. The three possible cases are illustrated in Sects. 2.3 and 2.4 with Esterel and Argos programs.

There are two distinct styles of synchronous modelling [11], which emerged during the development of the synchronous languages. The first one is known as *State Based Formalisms* (SBF), the second one is known as *Multiple Clocked Recurrent Systems* (MCRS's). The oldest and most developed synchronous language Esterel, uses the first style. Argos is also an SBF-style synchronous language. On the other hand, declarative dataflow languages Lustre and Signal use the second style.

State based formalisms are convenient for specifying control-dominated systems but they are not efficient in dataflow modelling. It is the opposite with MCRS's. Their main use is in specifying signal processing systems but it is more difficult to specify systems that step through different states. There have been attempts to unify two styles in a single environment as in [56, 57].

Synchronous programs can always be compiled into finite state machines. This property is very important since it greatly facilitates formal verification and ensures that memory requirements are known at compile time.

In Kahn process networks and related dataflow models, events are partially ordered. Events on a single channel are totally ordered, but they in general have no relation with events on other channels. In synchronous models, events are totally ordered. This has an important impact in modelling reactive systems, which have to promptly respond to every event that comes from the external environment. Reactive systems often have to wait for several events simultaneously. A KPN process cannot wait on multiple channels at the same time since it must implement blocking reads in order to achieve determinism. On the other hand, a synchronous process can test a channel before reading it and still preserve determinism. The downside of the total ordering of events is that it may unnecessarily reduce the implementation space by overspecifying the system, especially in the case of data-dominated systems.

It is interesting to note that due to the differences in event ordering, synchronous dataflow (SDF) is not an appropriate name when KPN based dataflow models are compared against synchronous models. To avoid confusion, a better name would be statically scheduled dataflow (SSDF) as suggested in [6].

2.5 Discrete Event Model

Discrete event (DE) [9] is the only MoC that incorporates the notion of physical time. Every event in DE carries a value and a time stamp. A DE block is activated when it receives an event to process. Events are processed chronologically. It is the

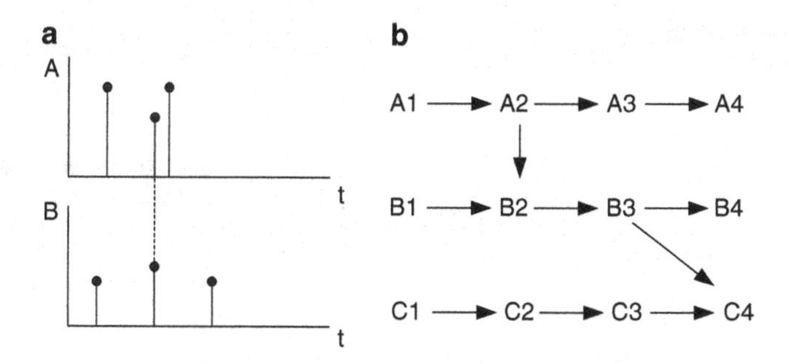

Fig. 2.7 Event ordering in DE and process networks. (**a**) Discrete event (**b**) Kahn process network

Fig. 2.8 Simultaneous events in DE

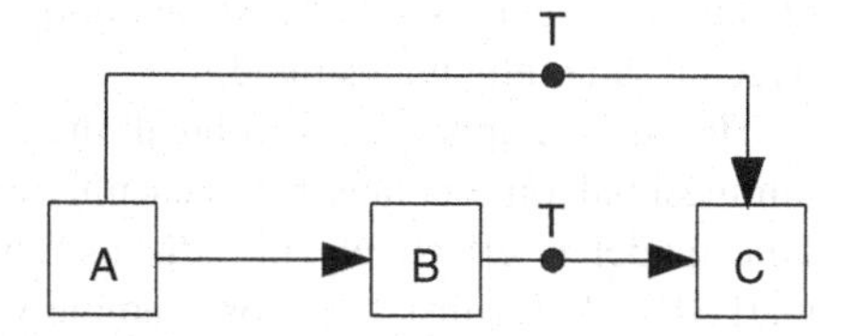

task of a DE scheduler to ensure that events with the smallest time stamp are processed first.

Events in DE are globally ordered. Figure 2.7 illustrates the difference in terms of ordering of events between the DE model and Kahn process networks.

In the DE model any two events are comparable even if they belong to different signals. Two events can either be simultaneous or one occurs before the other. This is shown in Fig. 2.7a. When a model employs a partial ordering scheme, as in Kahn process networks for instance, events that belong to the same signal are totally ordered but events across different signals may not be comparable at all. This is illustrated in Fig. 2.7b. There are three signals A, B and C. For example, events A1 and C1 are not related.

Total ordering of events can overspecify systems, which makes implementation more difficult. It is easier to build parallel systems when events across signals are not related. In theory the DE model should be suitable for modelling distributed systems, but creating a DE simulator for distributed systems can be a difficult task since there may be a very large number of events that need to be sorted.

It was mentioned in the previous section that events in SR are also totally ordered. However, the total ordering of events is easier to implement in SR than in DE, since every event in SR is related to the global clock. The SR compilers rely on the global clock to sort events.

Simultaneous events and feedback loops with zero delay are the two major problems in DE model. Figure 2.8 illustrates the problem with simultaneous events. Block B produces zero delay. The time stamp of an event that passes through block B remains unchanged. Therefore block C receives two events with the identical time stamp T, one from block A, and another from block B. The DE scheduler has to

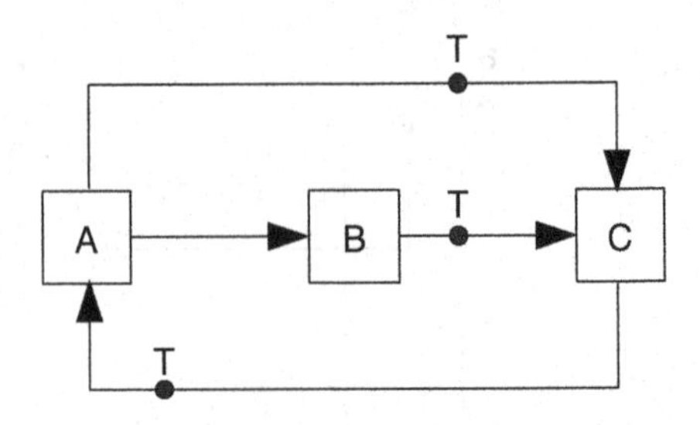

Fig. 2.9 Instantaneous feedback in DE

determine which event block C should process first or whether it should take two events at the same time, in which case both events are covered in one firing.

There are three different methods that the current DE simulators use in dealing with this problem. The ambiguity in Fig. 2.8 may be left unresolved. The processing of the two events is scheduled randomly. The result is a nondeterministic system which is generally undesirable.

The second approach is to break the relevant time instant into microsteps with infinitesimal time delays between microsteps. By doing that, a two-dimensional time model is effectively introduced. The time between two microsteps can be marked as Δt for instance. For example when an event passes through block B its time stamp is increased by Δt. The event appears at the output of block B with the time stamp $T+\Delta t$. When block C is invoked it processes first the event from block A which has the time stamp t. It then fires again and accepts the event from block B with the time stamp $T+\Delta t$. The infinitesimal time delays are not visible to the user. They are only used internally by the DE simulator.

The third approach is based on the analysis of data precedences statically within a single time instant, which is done in the Ptolemy [20] DE simulator. Arcs in a DE graph are assigned different priorities. The order of actions is always known when two simultaneous events appear at the same input.

Adding a feedback loop from block C to block A with C having zero processing time would create a situation that could not be resolved. This is shown in Fig. 2.9. An event circulates around the instantaneous loop without any increment in time stamp.

Digital hardware systems can be well described in DE. Both VHDL and Verilog simulators use DE as the underlying model of computation.

DE is mainly used for simulation. Global ordering of events with time stamps is expensive in real implementations.

2.6 Communicating Sequential Processes

A system described in communicating sequential processes (CSP) [13] consists of processes that execute independently in a sequential manner, but have to synchronize when they are communicating. When a process reaches a point of synchronisation with another process it has to stop and wait for the other process regardless of whether it has to read or write. Both reads and writes are blocking. Processes have to participate in common operations at the same time. This type of communication where processes have to synchronize in order to communicate is called *rendezvous*.

Every process is defined with its alphabet, which contains the names of events that are relevant to the description of the process. The actual occurrence of an event is regarded as instantaneous or atomic. When an activity with some duration needs to be represented, it should have a starting event and an ending event.

In [13] processes that describe various vending machines are used as examples. For example, a simple vending machine has only two relevant events – coin and choc (chocolate). When the machine gets a coin it gives out a chocolate. Generally upper case letters are used for processes and lower case letters are used for events. Let the process be called VM. Then its alphabet is written as

$$\alpha \mathrm{VM} = \{\mathrm{coin,\ choc}\}\ \alpha \text{ is used to denote an alphabet}$$

In order to explain the behaviour of CSP, several key CSP operators are introduced, together with examples, in the following paragraphs. Some operators that are less frequently used are omitted. A detailed description can be found in [13].

The sequential behaviour of a process is described with the prefix operator "$\rightarrow$". Let x be an event and let P be a process. Then $x \rightarrow P$ (pronounced "x then P") describes an object that which first engages in the event x and behaves as described by P. The prefix operator cannot be used between two processes. It would be incorrect to write $P \rightarrow Q$.

For example a simple vending machine that breaks after receiving a coin is described as:

$$\mathrm{VM1} = \mathrm{coin} \rightarrow \mathrm{STOP}$$

STOP is a process that indicates unsuccessful termination. STOP does not react to any events. A successful termination also exists in CSP and will be introduced later. If a process gets into that state, no event can occur. A simple vending machine that serves two customers before it breaks is described as follows:

$$\mathrm{VM2} = \mathrm{coin} \rightarrow \mathrm{choc} \rightarrow \mathrm{coin} \rightarrow \mathrm{choc} \rightarrow \mathrm{STOP}$$

The description indicates that the event choc can happen only after the event coin. The machine will not give a chocolate unless a coin is inserted.

Many processes will never stop. After executing a certain action they will go back to their initial state. Processes of that kind can best be described recursively. CSP supports recursive definitions, i.e. definitions in which a process name appears on both sides of the equation. For example the process CLOCK has only one event called tick. Thus the alphabet of the clock is $\alpha\mathrm{CLOCK} = \{\mathrm{tick}\}$. The process is recursively defined as

$$\mathrm{CLOCK} = \mathrm{tick} \rightarrow \mathrm{CLOCK}$$

The above definition is equivalent to $\mathrm{CLOCK} = \mathrm{tick} \rightarrow \mathrm{tick} \rightarrow \mathrm{CLOCK}$,

CLOCK=tick→tick→tick→CLOCK etc. This sequence can be unfolded as many times as necessary. Obviously recursive definitions are very useful in process descriptions.

Hoare seems to prefer another form of recursive definition that is more formal. For example a good vending machine that does not break is defined recursively in the way shown above as follows:

$$\text{VM3} = \text{coin} \rightarrow \text{choc} \rightarrow \text{VM3}$$

The alternative and more formal definition according to Hoare is in the form of μX:A.F(X) where the letter μ is used to denote a recursive expression, X is a local variable used in the recursive expression, and A is the alphabet of the expression i.e. the set of the names of events that appear in the expression. The alphabet is often omitted. Instead of X any other letter can be used, for example Y etc. VM3 is alternatively defined by μ with the definition below:

$$\text{VM3} = \mu X : \{\text{coin, choc}\}.(\text{coin} \rightarrow \text{choc} \rightarrow X)$$

Another important operator in CSP is the choice operator written as the bar |. The choice operator allows the environment in which the process operates to choose a sequence of actions that the process should perform. For example a vending machine may offer a choice of slots for inserting a 2p coin or a 1p coin. A customer decides which slot to use. The choice operator is used in conjunction with the prefix operator. In the expression below, two distinct events x and y initiate two distinct streams of behaviour:

$$(x \rightarrow P \mid y \rightarrow Q)$$

If x occurs before y, the subsequent behaviour of the object is defined by P. Similarly if y occurs before x, the subsequent behaviour of the object is defined by Q. The environment decides which event occurs first and thus which path is taken. The choice operator has to be used with the prefix operator. It would be incorrect to write P | Q.

As an example for the choice operator, a vending machine that offers a choice to the customer is defined below. The customer inserts a 5p coin and then chooses which combination of change to take.

$$\begin{aligned}\text{VM4} = \text{in5p} \rightarrow (&\text{out1p} \rightarrow \text{out1p} \rightarrow \text{out1p} \rightarrow \text{out2p} \rightarrow \text{VM4}\\ &\mid \text{out2p} \rightarrow \text{out1p} \rightarrow \text{out2p} \rightarrow \text{VM4})\end{aligned}$$

In all of the examples above only single processes were considered. In CSP processes can be made to run in parallel by using the concurrency operator ||. P || Q means that the system is composed of two processes that are running concurrently. The two processes have to synchronize on any event that is common to their alphabets. Events that are not common are executed independently by one of the two processes.

An interesting example that illustrates the use of the concurrency operator is given in [13]. Two processes are defined and then composed into a system using ||. The two processes are called NOISYVM (noisy vending machine) and CUST (customer). The alphabet of NOISYVM is defined below.

$$\alpha\text{NOISYVM} = \{\text{coin, choc, clink, clunk, toffee}\}$$

The machine offers chocolate or toffee. The event clink is the sound that a coin makes when it is inserted. The event clunk is another sound that the machine makes on completion of a transaction. This time the machine has run out of toffee and it only outputs a chocolate after receiving a coin.

$$\text{NOISYVM} = (\text{coin} \rightarrow \text{clink} \rightarrow \text{choc} \rightarrow \text{clunk} \rightarrow \text{NOISYVM})$$

The customer prefers to get toffee. When he doesn't get toffee he says a curse, which is included in the alphabet of CUST given below.

$$\alpha\text{CUST} = \{\text{coin, choc, curse, toffee}\}$$

$$\text{CUST} = (\text{coin} \rightarrow (\text{toffee} \rightarrow \text{CUST} \mid \text{curse} \rightarrow \text{choc} \rightarrow \text{CUST}))$$

The system that results from the concurrent composition of the two processes is defined below.

$$(\text{NOISYVM} \parallel \text{CUST}) = \mu X.(\text{coin} \rightarrow (\text{clink} \rightarrow \text{curse} \rightarrow \text{choc} \rightarrow \text{clunk} \rightarrow X \mid \text{curse} \rightarrow \text{clink} \rightarrow \text{choc} \rightarrow \text{clunk} \rightarrow X))$$

It should be noted that the two processes are synchronized on the event choc: the machine outputs it and the customer takes it. Choc appears in both alphabets. On the other hand, the events curse and clink occur asynchronously. They can occur one before the other or simultaneously. If they occur simultaneously it does not matter which one is recorded first.

This example clearly illustrates the essence of CSP: processes have to synchronize only on common events. Otherwise, they are asynchronous. Therefore CSP is not a completely synchronous MoC like synchronous/reactive model. In the SR model all actions are executed in lock-step. In CSP, processes synchronize only at rendezvous points.

Several other operators are briefly mentioned in the rest of the section.

When two processes P and Q are composed into a system, but do not have to synchronize on any events, their composition is written as P ||| Q which is read "P interleave Q". The events of the two processes are arbitrarily interleaved.

The unsuccessful termination is represented with the process STOP. The successful termination also exists and is represented by the process SKIP. SKIP does not react to any events in the same way as STOP. If a process finishes with SKIP it may be followed by another process. This is written as P;Q. Q is started after P successfully terminates.

In the examples shown so far, events could not be classified as inputs or outputs. In CSP the distinction between inputs and outputs is made using *channels*. Channels carry events. For example a simple process that inputs a message and then outputs the same message is defined as follows

$$\text{COPYBIT} = \mu X.(in?x \rightarrow (out!x \rightarrow X))$$

where $$\alpha in(\text{COPYBIT}) = aout(\text{COPYBIT}) = \{0,1\}$$

The alphabet of the input and output channels shows that the events 0 and 1 can occur on both channels.

Around the same time CSP was being developed, another similar model was emerging. The model is called Calculus of Communicating Systems (CCS). The model was created by Milner who later wrote a book about it [14]. Generally the two models are fairly similar, partly because the developers influenced each other while working on them. The basis of CCS are also processes that independently operate but have to synchronize on common events. Both models had a large impact on the research in concurrent systems. One of the main reasons for that is a sound formal treatment behind both models. An important property of concurrent systems such as deadlock can be formally analysed in CSP and CCS. Another example of formal analysis available in CSP and CCS is determining whether an implementation of a process satisfies its specification.

2.7 Petri Nets

Petri nets [12] is a graphical model that emerged in 1960s. Since then, many applications have been modeled with Petri nets and many research papers related to them have been published. While Petri nets is a tool for graphical modelling, it can also be mathematically analysed. In this section, though, only the basic features of Petri nets are introduced through examples.

Petri nets can be used to describe a wide range of applications. Systems that are suitable to be described by Petri nets are asynchronous, concurrent, distributed, and nondeterministic.

A Petri net is a bipartite directed graph. It consists of two kinds of nodes: places and transitions. Places are usually represented as circles while transitions are usually represented as bars or boxes. Places and transitions are joined by arcs. An arc can be drawn between a place and a transition but it is not allowed to join two places or two transitions. An arc can go from a transition to a place in which case it is marked as (t_i, p_i) or it can go from a place to a transition in which case it is marked as (p_i, t_i). It cannot go from a transition to a transition (t_i, t_j) or from a place to a place (p_i, p_j). With respect to a particular transition, a place can either be *input place* if the direction of the arc is (p,t) or *output place* if the direction of the arc is (t,p).

Places hold one or more tokens. Tokens are usually marked as dots inside places. When a Petri net *fires*, i.e. a transition is made, the numbers of tokens in various

A Petri net is a 5-tuple, $PN = (P, T, F, W, M_0)$ where:

$P = \{p_1, p_2, ..., p_m\}$ is a finite set of places,

$T = \{t_1, t_2, ..., t_n\}$ is a finite set of transitions

$F \subseteq (P \times T) \cup (T \times P)$ is a set of arcs (flow relation)

$W: F \rightarrow \{1,2,3, ...\}$ is a weight function

$M_0: P \rightarrow \{0,1,2,3, ...\}$ is the initial marking

$P \cap T = \emptyset$ and $P \cup T \neq \emptyset$

A Petri net structure PN = (P, T, F, W) without any specific initial marking is denoted by N.

A Petri net with the given initial marking is denoted by (N, M_0).

Fig. 2.10 Formal definition of Petri net

places change. The numbers of tokens in places represent a state of a Petri net which is in Petri nets terminology called *marking*. Each place is marked with the number of tokens it currently holds. The marking of a place is labelled as $M(p_i)$. The initial marking of a Petri net is denoted as M_0.

Arcs are weighted. The weighting of an arc shows how many tokens can flow through the arc during a transition. The weighting of an arc is labelled as w(p,t) or w(t,p) depending on the direction of an arc. Everything that has been said so far about Petri net can be summarized in the formal definition in Fig. 2.10.

The behaviour of a Petri net can be analyzed by observing the set of states (markings) that a Petri net steps through. A state change occurs upon a transition when markings of places (numbers of tokens they hold) are changed. The rule regarding transitions is described in the three points below:

- A transition is *enabled* in each input place is marked with at least w(p,t) tokens where w(p,t) is the weighting of the corresponding arc.
- An enabled transition may or may not occur.
- When a transition occurs w(p,t) tokens are removed from each input place and w(t,p) tokens are added to each output place.

A transition that has no input place is called *source* transition. A source transition is unconditionally enabled. A transition that has no output place is called *sink* transition. A sink transition consumes tokens but it does not produce any.

A transition p and a place t are called *self-loop* if p is both the input and output place of t. A Petri net that does not contain any self-loops is called *pure*. A Petri net is called *ordinary* if the weighting of each of its arcs is one.

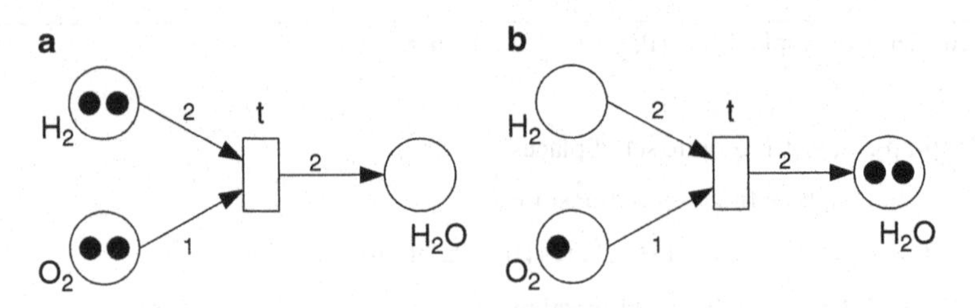

Fig. 2.11 Petri Net modeling chemical reaction. (**a**) Before transition (**b**) after transition

A simple example in Fig. 2.11 illustrates the transition rule. The Petri net in Fig. 2.11 models the well known chemical reaction $2H_2+O_2 \rightarrow 2\,H_2O$. Part (a) shows that transition t is enabled since each of the input places is marked with enough input tokens. The initial marking of the net is $M_0=(M(H_2)\ M(O_2)\ M(H_2O))=(2\ 2\ 0)$. Part (b) shows the state of the net after the transition has occurred. Two tokens have been removed from the input place H_2 as the weighting of the arc (H_2, t) is two. One token has been removed from the input place O_2 as the weighting of the arc (O_2, t) is one. Two tokens have been added to the output place H_2O as the weighting of the arc (t, H_2O) is two. After the transition the new marking is $M_1=(0\ 1\ 2)$. In part (b) the transition t is disabled since the place H_2 has no tokens and it has to have at least two.

The Petri net in Fig. 2.11 has an *infinite capacity*, because there is no bound on the number of tokens in each place. In practice, though, it may be more realistic to put a limit on the number of tokens in each place. Petri nets whose places are bounded are called *finite capacity* Petri nets. The capacity of place p is labelled as K(p).

When a Petri net is a finite capacity net, a transition has to account for the finite capacity of places. After a transition, the number of tokens in the output places must not exceed their capacities. This kind of transition rule that takes into account the capacity constraints is called *strong* transition rule. In the above example, *weak* transition rule was applied since the Petri net was an infinite capacity net, so the capacity constrains were not relevant.

In fact there are two options that can be used on a finite capacity net denoted as (N, M_0). Either the strong transition rule can be used or the weak transition rule can be used on the transformed net (N',M_0'). The transformation from (N, M_0) to (N',M_0') consists of the two steps shown below:

- For each place p in (N, M_0) add a complementary place in (N',M_0'). The initial marking of the complementary place should be $M_0'(p')=K(p)-M_0(p)$.
- New arcs must be added to connect complementary places to transitions. For every arc (t,p) add a new arc (p′,t) with weighting w(p′,t) = w(t,p). For every arc (p,t) add a new arc (t, p′) with weighting w(t, p′) = w(p,t).

The effect of this transformation is that the sum of tokens in a place p and its complementary place p′ is equal to the capacity of the place p (K(p)) before and after transition when the weak transition rule is applied. Figure 2.12 gives an example of the transformation.

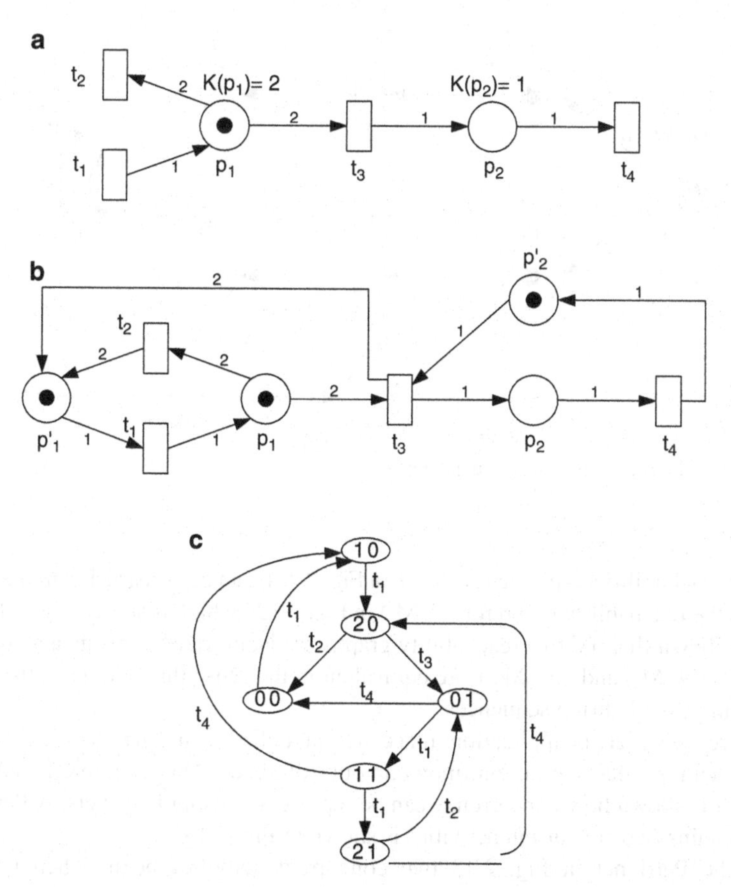

Fig. 2.12 Finite capacity net, its transformation and reachability graph (**a**) Finite capacity Petri net (**b**) finite capacity net from (**a**) transformed (**c**) reachability graph for Petri net in (**a**)

Part (a) shows a finite capacity net. The initial marking is $M_0=(1\ 0)$ and the only enabled transition is t_1. After t_1 fires the new marking is $M_1=(2\ 0)$. The transitions t_2 and t_3 are now enabled. If t_2 fires the new marking is $M_2=(0\ 0)$. Otherwise if t_3 fires the new marking is $M_3=(0\ 1)$. By repeating this process a reachability graph can be drawn which is shown in Fig. 2.12c.

The finite capacity net (N, M_0) in part (a) can be transformed into (N', M_0') shown in part (b). The first step is to add complementary places p_1' and p_2' with their initial markings $M_0'(p_1')=K(p_1)-M_0(p_1)=2-1=1$ and $M_0'(p_2')=K(p_2)-M_0(p_2)=1-0=1$. The second step is to add new arcs to connect the complementary places to the transitions. For example for p_2', (t_4, p_2') is added with $w(t_4, p_2')=1$ since $w(p_2, t_4)=1$ and (p_2', t_3) with $w(p_2', t_3)=1$ since $w(t_3, p_2)=1$. Similarly for p_1' three arc are drawn: (t_2, p_1'), (p_1', t_1) and (t_3, p_1') with $w(t_2, p_1')=w(p_1, t_2)=2$, $w(p_1', t_1)=w(t_1, p_1)=1$ and $w(t_3, p_1')=w(p_1, t_3)=2$.

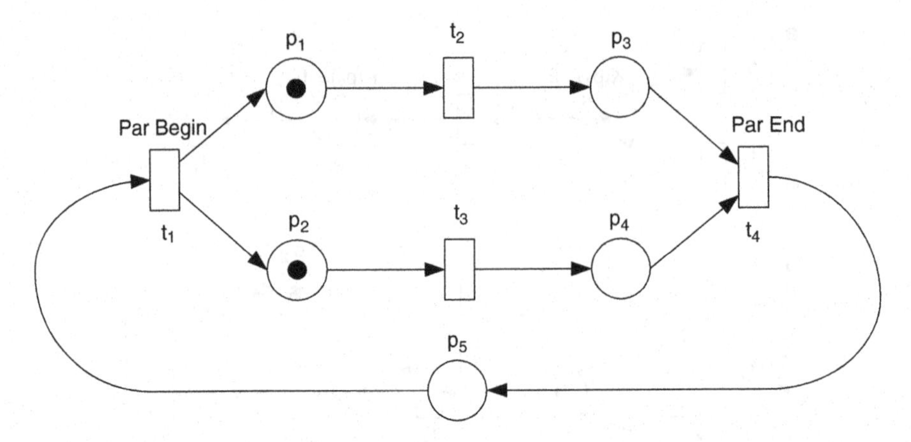

Fig. 2.13 A Petri net with concurrent activities

The reachability graph for (N', M_0') in Fig. 2.12b can be constructed in the same way as the reachability graph for (N, M_0) in Fig. 2.12a which is shown Fig. 2.12c. It can be shown that the two reachability graphs are isomorphic. This means that the two nets (N, M_0) and (N', M_0') are equivalent in the sense that both have the same set of all possible firing sequences.

There is a variety of applications that can be modelled with Petri nets. Considering all of them would take an enormous amount of space. One example is selected where it is shown how concurrency can be represented using Petri nets. A Petri net that contains two concurrent activities is shown in Fig. 2.13.

In the Petri net in Fig. 2.13 two concurrent activities begin when t_1 fires. Transitions t_2 and t_3 can fire independently of each other. This is enabled by the fact that all places in the two parallel branches have only one incoming and one outgoing arc (p_5 is also like that). The Petri net in Fig. 2.13 belongs to the subclass of Petri nets called *marked graph*. In a marked graph all places have exactly one incoming and one outgoing arc.

In embedded systems applications a clear disadvantage of Petri nets is the lack of hierarchy. Also, Petri nets are highly nondeterministic because of nondeterministic firings of transitions. Nondeterminism can make the analysis of a large embedded system difficult. It has been shown, however, that many smaller embedded applications can be successfully represented with Petri nets.

2.8 Statecharts/Statemate

It was pointed out in Sect. 2.1 that flat sequential FSMs are inadequate for representing complex control systems. The number of states and transitions becomes very large and the whole system becomes unmanageable. Harel introduced in [22]

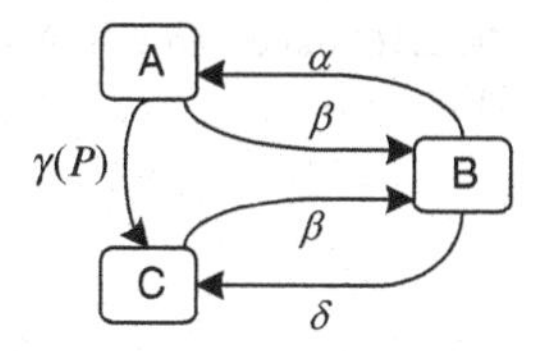

Fig. 2.14 A basic FSM

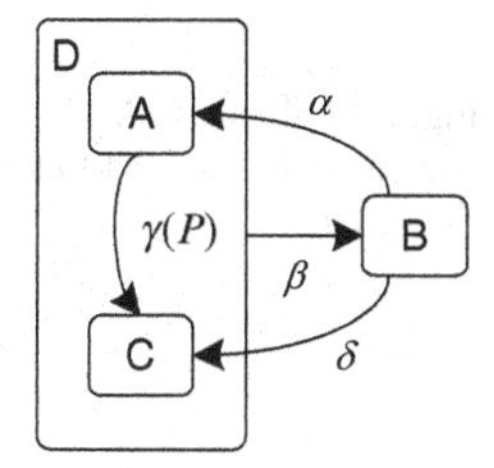

Fig. 2.15 Hierarchical equivalent of Fig. 2.14

some innovations that largely improved the usefulness of FSMs. The resulting visual language was called Statecharts. A tool called Statemate based around Statecharts was also developed. Both are discussed in this section.

In Harel's paper an FSM transition is labelled as *a[b]/c*. This type of notation was already introduced in Sect. 2.1. *a* is the event that causes the transition, *b* represents the condition which has to be fulfilled for the transition to take place when *a* occurs, and *c* is the output event generated by the transition which causes other transitions as will be illustrated later in the section. In Statecharts, the output events are called *actions*.

The basic FSM is neither hierarchical nor concurrent. Hierarchy and concurrency are common features in embedded systems. In Statecharts the two key innovations are OR states used to describe hierarchy and AND states used to describe concurrency. Figures 2.14 and 2.15 illustrate the use of OR states. Figure 2.14 shows a basic flat FSM with three states A, B, C. Figure 2.15 shows an equivalent hierarchical FSM.

In the FSM in Fig. 2.14 the same event β leads from states A and C to state B. It is convenient to cluster these two states into a *superstate* called D in Fig. 2.15. State D contains two *OR* states A and C. When D is entered, either A or C becomes active but not both. So OR states actually behave as exclusive OR states. When the event β occurs D is left which means that either A or C is left, and B is entered. It should be noticed that in Fig. 2.14 two arcs are used to represent this change while in Fig. 2.15 only one arc is used. In this simple example the FSM in Fig. 2.14 can be easily understood, and reducing the number of transitions may not seem significant, but in complex systems reducing the number of transitions brings large benefits. By introducing hierarchy with OR states the number of transitions is reduced and it is generally easier to understand the system.

The second key innovation in Statecharts is the use of AND states to represent concurrency. Figure 2.16 shows how AND states are represented. State Y is an

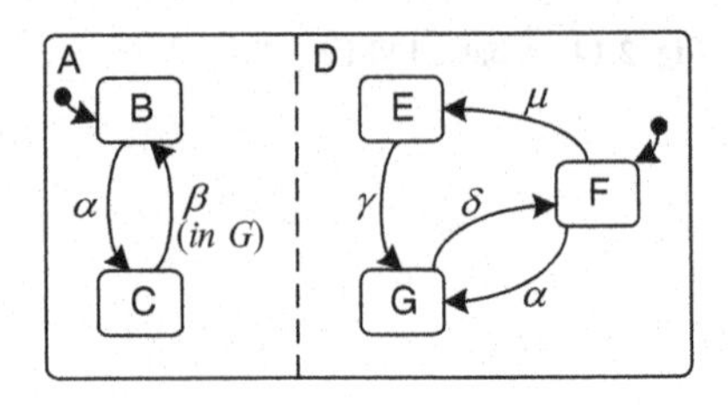

Fig. 2.16 Statechart with AND states

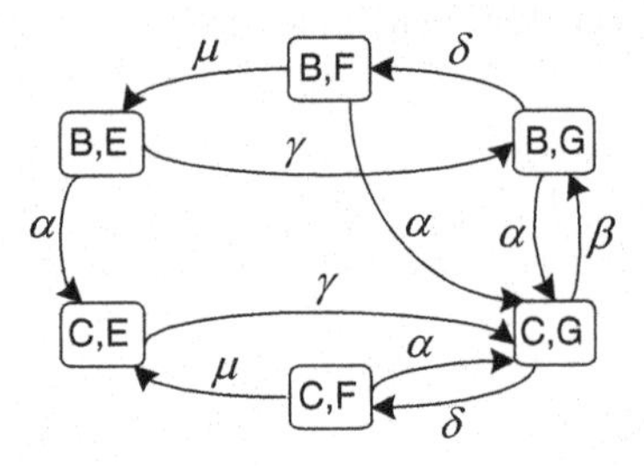

Fig. 2.17 Basic FSM equivalent to statechart in Fig. 2.16

orthogonal product of states A and D. *Orhogonality* is the term that is often used in [22] to denote concurrency. Being in state Y means that both states A and D are simultaneously active. Thus A and D are AND states. Graphically AND states are separated by a dashed line.

State A contains OR states B and C, while state D contains OR states F, E and G. Arrows pointed to B and F mean that these two states are the default states for A and D, respectively. This means that, when Y is entered from outside, states B and F become simultaneously active. Simultaneous transitions are possible in AND states. For example if B and F are active and the event α occurs states C and G are entered at the same time. Some events can cause a change in only one AND state. For example if F is active and μ occurs, E is entered but there is no change in A. So while AND states are concurrent they can operate independently.

The equivalent basic FSM of the statechart in Fig. 2.16 is shown in Fig. 2.17. The FSM in Fig. 2.17 contains six states since in Fig. 2.16 A contains two states and D contains three states. If A and D contained thousand states each the FSM in Fig. 2.17 would contain million states. This is so called state explosion problem, which makes it very difficult to represent large concurrent systems with basic FSMs.

There are some other features of Statecharts such as history, condition and selection entrances, timeouts and delays, which are not discussed here. Those features are less important than OR and AND states described above.

In the examples so far, FSM transitions did not produce any actions. Those transitions were in the form of *a[b]* rather than *a[b]/c*. Actions can trigger other transitions so they are not really different from other events. In Statecharts, concurrent states communicate by the broadcast mechanism. Actions are instantaneous and they are immediately visible to all states. A state does not have to be prepared to receive an action as in rendezvous.

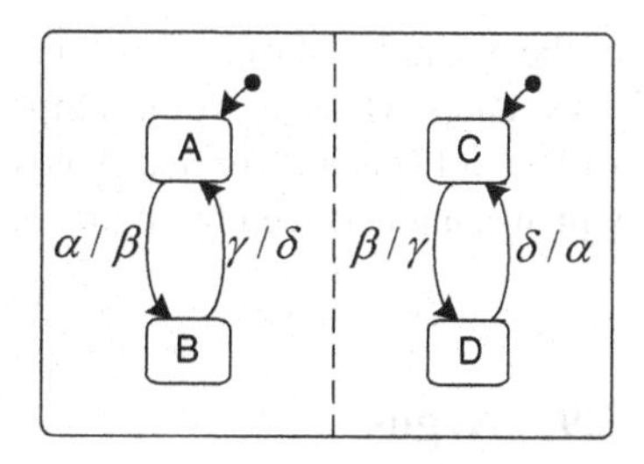

Fig. 2.18 Infinite loop due to instantaneous broadcast

The broadcast mechanism can create some interesting semantic problems. Harel outlined in [22] some of those problems and did not provide an immediate solution. That triggered lots of research on Statecharts resulting in at least 20 different variants. Von der Beeck provides a good summary of Statecharts variants in [68]. Figure 2.18 illustrates one of the problems related to communication in Statecharts.

It is not difficult to see that an infinite loop of transitions is produced when any of the four events occurs. A tool that analyses Statecharts must be able to detect directed cycles like this and report an error.

Statecharts can represent the reactive part of an embedded system but they cannot be used for transformative data functions. Statemate [82] is a tool based around Statecharts that allows specification and analysis of a complete embedded system including data parts.

Statemate incorporates *module-charts, activity-charts and statecharts*. In the examples above it was shown how an action produced in a transition can cause other transitions. In Statemate actions can also trigger *activities*. Unlike actions, activities take time. Activities are represented by activity-charts. The purpose of activity-charts is to represent data parts of the system under development (SUD) in Statemate. SUD is the term used in [82], a paper that describes Statemate.

In Statemate, statecharts are used to represent the reactive part of a system, but they are also used to control activity-charts, which, represent the data part of a system. Module-charts represent statecharts and activity-charts at a level that is closer to physical implementation. For example module-charts indicate how an activity-chart maps to a particular processor.

Statecharts control activity-charts by actions produced upon state transitions. Examples of some actions are start(a), suspend(a), resume(a), stop(a) where *a* is an activity. Statecharts and activity charts can be mixed at any level of hierarchy.

Atomic activities (those that cannot be further decomposed) are described by a programming language, but it is not indicated in [82] which programming language has to be used. Perhaps, several different programming languages may be used.

After a system has been described in Statemate using module-charts, activity-charts and statecharts, it can be simulated. Statemate can also synthesize the system description into C-code.

Although Statemate is equipped to support a complete design of embedded systems, statecharts that are used to model reactive behaviour are definitely the part that attracted the most attention. It is interesting to note that in Statecharts it is possible for an arc to connect two states at different hierarchical levels as seen in some of the examples above. This jumping across hierarchy has been criticized by some researchers who believe it compromises the modularity of a design. In fact some variants of Statecharts like Argos allow declaration of local signals inside a state, which cannot be seen by higher level states.

2.9 Argos

A single, flat FSM is suitable for specifying sequential controllers, but it can hardly handle more complicated embedded systems where support for concurrency and hierarchy becomes necessary. This weakness was addressed in Statecharts [22], which introduced hierarchical and concurrent compositions of FSMs and several other useful features. It was described in [22] that concurrent FSMs communicate using synchronous broadcast with the assumption of instantaneous communication. However, many questions regarding instantaneous loops were left unanswered. This semantic gap resulted in more than 20 different versions of Statecharts [68] approaching causality problems in different ways. Argos is a version of Statecharts that uses the principles of Esterel to deal with semantic challenges caused by instantaneous communication. Unlike other Statecharts variants, Argos complier rejects all programs that contain non-determinism that the programmer did not intend to have, i.e. implicit non-determinism. Another distinguishing feature of Argos is that it forbids inter-level transitions, which connect two states on different hierarchical levels. This emphasizes modularity.

The three basic Argos operators that are applied on FSMs are: synchronous parallel, refinement and hiding (localization). They were introduced in [69]. Later, a few additional operators were described in [24]. Figure 2.19 shows an Argos specification with the three main operators.

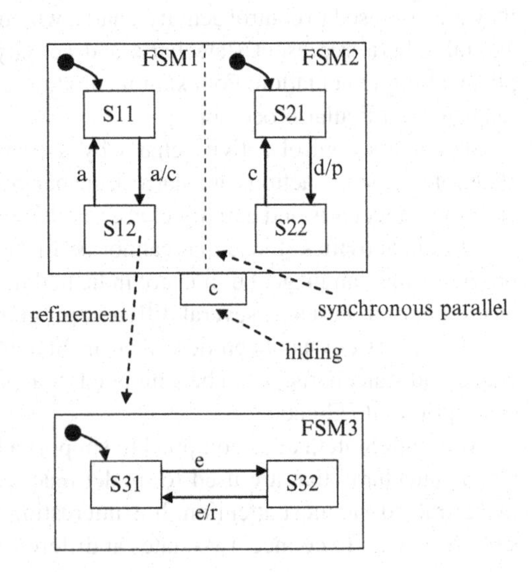

Fig. 2.19 Argos example

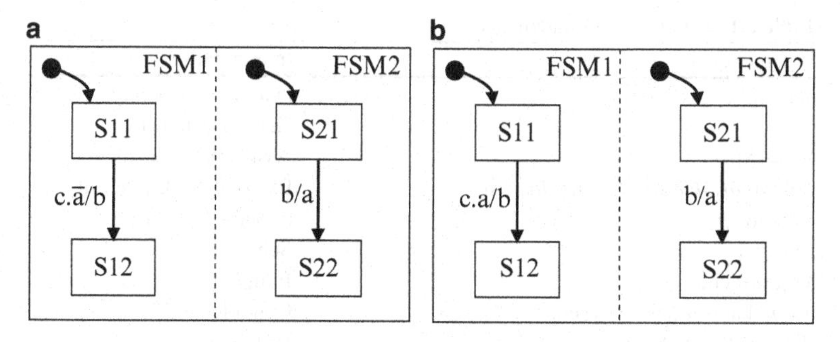

Fig. 2.20 Causality errors (**a**) no solution (**b**) multiple solutions

The top level FSMs are connected by the synchronous parallel operator and communicate using the local signal *c*. When FSM1 is in S11 and *a* is present, it makes the transition to S12 and emits *c*. If FSM2 is in S22, the presence of *c* instantaneously triggers the transition to S21. When FSM1 makes the transition from S12 to S11 it preempts FSM3. Only weak abort is available in Argos. Thus FSM3 is allowed to react in the tick in which it is pre-empted. For example if it is in S32 and *e* is present in the instant of pre-emption, *r* will be emitted.

As in Esterel, causality problems due to instantaneous communication can also appear in Argos. Two examples are shown in Fig. 2.20. A dot between input signals means logical AND (conjunction). A line above the name of a signal denotes its absence.

The specification in Fig. 2.20a has no meaningful behaviour. If *c* is present and *a* is absent, FSM1 makes the transition from S11 to S12 and emits *b*. Since *b* is present FSM2 makes the transition from S21 to S22 and emits *a*. This means that *a* is present and absent simultaneously, which is not possible. Figure 2.20b illustrates non-determinism. If *c* is present, both FSMs can take transitions and emit signals *a* and *b* that are necessary for their activations or transitions are not taken and *a* and *b* are not emitted.

2.10 Esterel

Esterel is an imperative language based on the synchronous reactive model of computation that was described in Sect. 2.3. Since the synchrony hypothesis is used in Esterel, it is assumed that all communications and computations take zero time. As a result outputs are synchronous to inputs. Communication is achieved by synchronous broadcast. Events produced by one process are immediately visible to other processes. Esterel is a deterministic language. For a given sequence of inputs only one sequence of outputs is possible.

Esterel programs communicate with external environment through signals and sensors. Signals are used as inputs and outputs. Sensors can only be used as inputs.

Table 2.1 Esterel kernel statements

Statement	Interpretation
nothing	Dummy statement
halt	Halting statement
X : = exp	Assignment statement
call P (*variable-list*) (*expression-list*)	External procedure call
emit S(*exp*)	Signal emission
$stat_1$; $stat_2$	Sequence
loop *stat* end	Infinite loop
if *exp* than $stat_1$ else $stat_2$ end	Conditional
abort *stat* when S	Watchdog
$stat_1$ || $stat_2$	Parallel statement
trap T in *stat* end	Trap definition
exit T	Exit from trap
var X : *type* in *stat* end	Local variable declaration
signal S (combine *type* with *comb*) in *stat* end	Local signal declaration

Signals can be *pure* or *valued.* Pure signals only have status at each instant. The status of a signal indicates whether the signal is present or absent. Valued signals also carry a value besides status. Valued signal S is denoted as S(v) where *v* represents the value. In expressions, the value of the signal S is denoted as ?S. If different values of the same signal are produced by multiple processes at the same time instant, it has to be specified how those values are combined. For example different values can be added to produce the resulting value. It can also be specified that it is forbidden to have multiple values for a signal at any time instant.

There are two types of Esterel statements: *primitive* statements and *derived* statements. Derived statements are actually derived from primitive statements. From the point of view of the programmer, derived statements are user-friendly and they also make programs shorter. An example of how a statement is derived will be shown later in this section. Primitive statements can be divided in two groups: basic imperative statements and temporal statements. Basic imperative statements take no time, they are executed instantaneously. Temporal statements handle signals and they take time. Table 2.1 lists the primitive statements, which form the basic Esterel kernel. The table was taken from [15] slightly modified to reflect recent changes in the Esterel syntax.

Most of the statements in Table 2.1 are found in other imperative languages. The statements *emit* and *abort when* are specific to Esterel. *Nothing* does nothing and takes no time. *Halt* does nothing but it never terminates so it takes time. The assignment statement and external procedure call are both instantaneous. *Emit* evaluates the value of the signal, emits the signal with its value and terminates. The emission is instantaneous. The sequence of statements is executed sequentially but since the computation is assumed to be infinitely fast the whole sequence takes no time. For example, a sequence

$$X := 1; X := X + 1$$

results in X being equal to 2. The sequence above takes zero time, but it is executed in the correct order. The number of statements in a sequence has to be finite. For example statements such as

$$X := 0;\ \text{loop} X := X + 1 \text{end}$$

are not allowed. The loop construct never terminates. When the body of a loop terminates the loop is instantly restarted. The expression in *if then else* conditional can be composed of both signals and variables in the latest version of Esterel. Thus, the *present* statement which is used exclusively for testing signals became obsolete.

The execution of *abort when* can end in two ways. The body inside the construct can finish before signal S occurs. If S occurs before the body finishes, the body is terminated instantly without being allowed to execute in the instant of termination. This type of abort is *strong*. It is also possible to create a *weak* abort simply by writing *weak abort* instead of *abort*. Weak abort can also be made using *trap*. When weak abort is used, the body is allowed to execute in the instant of termination.

The branches of a parallel statement start simultaneously. A parallel statement terminates when all of its branches have terminated.

The *trap exit* construct is a powerful control mechanism also found in some other languages. *Trap* defines a block of statements that instantly terminate when the corresponding *exit* statement is executed.

There are several useful derived statements in Esterel such as *await, every, each, sustain*. It is shown below what *await* is equivalent to.

```
await S      instead of      abort
                             halt
                             when S
```

The basic unit of an Esterel program is called *module*. A module consists of the *declaration* part and *statement part*. In the declaration part, types, constants, functions and procedures used in the module are declared. Esterel is used in combination with a host language, which can be C for instance. Earlier Esterel versions had only basic types and operators built in, such as integer, Boolean, basic arithmetic and logic operators. More complex operators and types had to be imported from the host language. The latest Esterel version [58] includes more complicated data types that are mainly geared towards hardware modelling. The declaration part can also specify relations between signals. For example a relation can state that two signals never occur at the same time. This helps the Esterel compiler in optimisation. An example of an Esterel module is given in Fig. 2.21.

The first top level parallel branch emits O with the value of 1 after either A or B occurs. This operation can be pre-empted by R, which is emitted by the second parallel branch when C occurs.

It was mentioned in Sect. 2.3 that the powerful zero delay communication mechanism of the SR model can create causality problems. In particular, two kinds of problems were mentioned: lack of solution and multiple solutions.

Fig. 2.21 Esterel example

```
module est_example:

input A, B, R;
output O : integer;

loop
  abort
    [ await A || await B ];
    emit O(1)
  when R
end
||
loop
  await C;
  emit R
end

end module
```

Those two problems will now be illustrated on Esterel programs. The program below demonstrates the first problem – no solution.

```
signal S in
  present S else emit S end
end
```

If it is assumed that signal S is not present then it is emitted so it is present. This contradiction cannot be resolved since the whole statement is executed in a single instant. The second program demonstrates the second causality problem – multiple solutions.

```
signal S1, S2 in
  present S1 else emit S2 end
  ||
  present S2 else emit S1 end
end
```

The program has two solutions. In one solution S1 is present and S2 is absent. In the other solution S2 is present and S1 is absent. Hence the program is nondeterministic. An Esterel compiler has to reject both programs.

The first Esterel compiler [15] converts an Esterel program into a flat FSM. The resulting code is very fast but it is impractical for large systems due to exponential increase in code size. The second Esterel compiler [59] translates an Esterel program into a synchronous digital circuit. The resulting code is compact, but its execution is very slow since every gate has to be evaluated in every reaction. Recent Esterel compilers [60–62] produce fast and compact code by simulating the reactive features of Esterel, but they cannot compile a number of correct cyclic Esterel programs.

Esterel's reactive statements have inspired several modifications of C/C++ and Java to make them more suitable for embedded systems. Some examples are Modeling reactive systems in Java [63], ECL [64], Jester [65], Reactive-C[66], JavaTime [67].

2.11 Lustre and Signal

Lustre [36] and Signal [37] also belong to the group of synchronous languages. Unlike Esterel, which is an imperative language, Lustre and Signal are declarative dataflow languages. The main motivation behind the creation of Lustre and Signal was the fact that most embedded system designers have background in signal processing and control systems, not computer science. Signal processing/control systems are often modelled with equations in which variables are expressed as functions of time. Therefore, according to the creators of Lustre and Signal, dataflow languages would be a natural choice for control/signal processing engineers rather than imperative languages computer scientists are used to.

In Lustre and Signal, every variable refers to a *flow* which is a sequence of values associated with a clock. Lustre and Signal have more powerful clocking schemes than Esterel. For example, in Lustre, it is very easy to derive slower clocks using flows of Boolean values. This is illustrated in Table 2.2.

A clock derived from a Boolean flow represents a sequence of instants where the flow is true. The basic clock is the fastest clock. In the table above, the Boolean flow B1 is associated with the fastest clock. B2 is a Boolean flow that is associated with the clock derived from B1. A clock can also be derived from B2.

Synchronous dataflow languages can successfully describe some systems where clock rates are multiples of each other. It is more difficult to describe systems where data arrives irregularly. In that case, events that indicate absence of samples can be used, but in this way the system specification tends to become inefficient.

In [36] it is stated that Lustre can effectively describe both signal processing and reactive systems. Some examples that show how Lustre can describe reactive systems are given. It is true that Lustre can specify reactive systems, but it is not straightforward to do that. Synchronous dataflow is convenient for specifying some signal processing systems, but not reactive systems since states are not easily specified.

Table 2.2 Boolean flows and clocks in Lustre

Basic clock	1	2	3	4	5	6	7	8
Values of B1	True	True	False	True	False	True	False	False
Clock derived from B1	1	2		3		4		
Values of B2	False	True		True		False		
Clock derived from B2		1		2				

Synchronous dataflow languages are often regarded as being close to Kahn process networks. However there is one important difference – synchronous dataflow languages are synchronous, while Kahn process networks are asynchronous. Kahn process networks use buffers for communication between nodes. This is not the case in Lustre and Signal.

2.12 SystemC

SystemC [70–72] is an attempt to expand the widely used programming language C++ with features that would make it suitable for specifying embedded systems. SystemC is a class library of C++ and can be compiled on any C++ compiler. This is a clear advantage since there are well-developed, mature tools for C++. SystemC includes constructs to support specification of embedded systems characteristics such as timing, concurrency and reactive behaviour. Such construct cannot be found in the standard C/C++ programming environment.

SystemC enables designers to represent both hardware and software parts of an embedded system in the single environment. A specification in SystemC is executable because it can be simulated. In fact, SystemC allows specifying and simulating a system at different levels of abstraction, ranging from highly abstract system models down to cycle based models. The ability to simulate a high level system specification brings large benefits to the design process. System functionality can be verified before implementation begins. Mistakes can be uncovered early in the design process, when it is much cheaper to remove them than in well advanced design stages.

The main idea behind SystemC version 1.0 is to enable the designers to describe both software and hardware using the same language. For that purpose some features of hardware description languages were incorporated in the SystemC class library. Later SystemC 1.0 evolved into SystemC 2.0. The current version is 2.2 but we will focus only on major changes that occurred between the releases 1.0 and 2.0. With respect to SystemC 1.0, SystemC 2.0 is better equipped for system-level modelling. This is mainly due to new communication mechanisms that appeared in SystemC 2.0. The second version still supports everything from the first version. In the following paragraphs SystemC 1.0 is firstly described. Later in the section, the features that defined SystemC 2.0 are outlined.

A specification in SystemC consists of modules. A module can instantiate lower level modules, thus supporting hierarchy. Modules contain processes that describe system functionality. Three kinds of processes exist as will be described below. Modules are connected through ports which can be bi-directional or uni-directional. Ports are connected with signals. There is a wide range of signal types available in SystemC in order to support different levels of abstraction. Clocks are also available as a special signal type.

Breaking a system into modules enables division of tasks among designers. A module can be modified such that its external interface and functionality remain the same and the only thing that is changed is the way in which the function is described. In this way other modules in the design are unaffected. The external interface of a module is represented by its ports. There are three types of ports: input, output and input – output ports. Every port also has a data type associated with it. Modules are declared with the SystemC keyword SC_MODULE. For example the declaration of a module that describes a fifo (first-in-first-out) buffer is given below.

```
SC_MODULE (fifo) {
  sc_in<bool>load;
  sc_in<bool>read;
  sc_inout<int>data;
  sc_out<bool>full;
  sc_out<bool>empty;
  // module description not shown
}
```

Only the ports of the module are shown, not the functionality. The module has two input ports, two output ports and one bi-directional input-output port. The input and output ports are used for control and are all of type boolean. The bi-directional data port is of type integer.

Ports of different modules are connected with signals. A signal declaration only indicates which data type is carried by the signal, while the direction (in, out, inout) is not specified as it is in the case of ports. The direction of data flow in a signal is determined by ports that are connected together by the signal.

Local variables can be used to store data received through ports. Local variables are only visible inside the module in which they are declared, unless they are explicitly made to be visible in other modules. Local variables can be of any C++ data type, SystemC data type or user-defined data type.

The functionality in a module is defined by processes. Processes are registered with the SystemC kernel. Processes are sensitive to signal changes. A process starts executing when there is a change in at least one of the signals the process is sensitive to. Some processes are similar to C++ functions. They execute sequentially and return control once they have stepped through all statements. Other processes are different in that they may be suspended by halting statements and then resumed at some later instant.

There are three types of processes in SystemC: methods, threads and clocked threads. Methods behave like C++ functions. A method starts executing statements sequentially when called. It returns control to the calling mechanism when it reaches the end. The execution of a method has to be completed. For that reason, it is recommended in SystemC User's Guide that designers should be careful to avoid making infinite loops inside a method.

Threads may be suspended and resumed. A thread is suspended when the wait( ) function is encountered. It is resumed when one of the signals in its sensitivity list changes.

Clocked threads are a special case of threads. Clocked threads are suspended and resumed in the same way as threads. However the sensitivity list of a clocked thread contains only one signal, and that signal is a clock. Furthermore, a clocked thread is sensitive to only one edge of the clock. Clocked threads resemble the way in which synchronous digital circuits are specified. For example, a synchronous process in VHDL contains only a clock signal in its sensitivity list.

```
#include "systemc.h"

SC_MODULE (count) {
   sc_in <bool>     load;
   sc_in <int>      din;
   sc_in <bool>     clock;
   sc_out <int>     dout;

   int count_val;       // internal data storage

   void count_up( );

   SC_CTOR (count)   {
      SC_METHOD (count_up);              // Method process
      sensitive_pos   <<  clock;
   }
};

void count : : count_up ( ) {
   if (load)  {
        count_val = din;
   } else  {
       count_val = count_val +1;
   }
   dout = count_val;
}
```

Fig. 2.22 SystemC description of a counter

The code in Fig. 2.22 illustrates how a counter is specified in SystemC inside a module.

The behaviour of the module is described by the process count_up, which is of the type method. The process is triggered by the positive edge of the clock. This is specified in the line "sensitive_pos<< clock". When the process is triggered, the value on input port load is checked. If it is true, variable count_val is assigned the value of input port din. Otherwise, count_val is incremented by one. count_val is a local variable, visible only in the module. It should be noted that every module is initialized by a constructor. The keyword SC_CTOR is used for constructors.

In SystemC 1.0 modules communicate through ports which are connected by hardware signals. The behaviour of those signals is essentially the same as in VHDL and Verilog. From the point of view of the system level designer, communication using only hardware signals is insufficient to capture highly abstract features that are apparent at the system level. For that reason new communication mechanisms were added leading to the second release. For example, in SystemC 2.0, a semaphore or a mutex are available to protect shared data used by communicating modules. Modules can also communicate using FIFOs, another type of communication not directly supported in SystemC 1.0. Moreover, it is possible for users to define their own communication mechanisms. In order to do that, the user has to define a new *channel* and its *interfaces*. Channels and interfaces are two new kinds of objects in SystemC 2.0. A channel defines a communication mechanism. An interface is used to connect a port of a module to a channel. FIFO, mutex and semaphore are examples of built-in SystemC 2.0 channels.

With abstract communication mechanisms, it becomes possible to work at levels of abstraction that are higher than RTL. *Transaction level modelling* (TLM) [71] is an example of such a level. In TLM, processing elements typically communicate by procedure calls, whereby a whole packet of data can be transferred. Implementation details seen at the register transfer level are omitted. As a result, TLM simulations are much faster than RTL simulations. Extensive research in modelling embedded systems at the TLM level has been carried out recently [73–78].

Designs in SystemC can be simulated with testbenches. A testbench typically contains a module that generates inputs for the design under test (DUT), the DUT itself, and a module that checks the outputs of the DUT. It is often the case that a testbench does not contain a module for checking outputs; instead the designer manually checks outputs. Designs at various levels of abstractions can be simulated.

It is worth mentioning that the development of interfaces and channels in SystemC 2.0 was significantly influenced by another system-level language based on C, called SpecC [79]. While SystemC libraries are defined using the standard C/C++ syntax, SpecC extends the standard ANSI-C by adding new constructs to support hardware and system-level features. HardwareC [80] and Handel-C [81] are also examples of languages that extend C in order to support hardware design.

2.13 Ptolemy

Ptolemy is an environment for modelling heterogeneous systems. Embedded systems are often heterogeneous in that they encompass different models of computation. Their heterogeneity also stems from the fact that they are composed of software and hardware components.

The basic object in Ptolemy is called *block*. In the first version of Ptolemy [83] blocks are described in C++. In the second, expanded version [84] blocks are described in Java. Blocks communicate with the external environment through portholes. Blocks can use different methods for communication. Other objects in Ptolemy are derived from blocks. It should be noted that Ptolemy intensively uses object oriented programming.

Ptolemy has several domains. Domains represent models of computation. Examples of the domains in Ptolemy are process networks (PN), dynamic dataflow (DDF), Boolean data flow (BDF), synchronous dataflow (SDF), discrete event (DE), synchronous/reactive (SR). SDF is the most developed domain in Ptolemy. In fact, Ptolemy's predecessors only supported the dataflow model of computation. A *star* is an object derived from a block. A star always belongs to a certain domain. An object called *galaxy* can be formed from stars. Galaxies are hierarchical – they can contain other galaxies.

Another object derived from a block is called *target*. There are two types of targets in Ptolemy: simulation and code generation. An object called *scheduler* is needed to determine the order of execution of stars. Like stars, schedulers are also related to domains. Different domains have different schedulers. Finally, putting together stars, galaxies, schedulers and a target results in a complete application is called *universe*. What is done with a universe depends on its target – it can be either simulated or code can be generated from it (C and VHDL code generators are available).

The key feature of Ptolemy is that stars from different domains can be mixed. This allows multiple models of computation to be present in a single system. It is important to emphasize that one domain can be embedded inside another, hierarchically, but different domains cannot be at the same hierarchical level. A domain can be embedded inside another by the mechanism called *wormhole*. Externally, a wormhole behaves according to the semantics of the domain it is in, just like any other star or galaxy that belong to that domain. However, the internal behaviour of a wormhole is entirely different because it contains another domain with different semantics. The interface between the domain that resides in a wormhole and the external domain in which the wormhole is placed is called *Event Horizon*. Two conversions take place on the event horizon. One of them is the conversion between particles that cross the interface from one domain to another. For example, if DE (discrete event) is embedded inside SDF (synchronous data flow), a particle going from SDF to DE has to be attached a time stamp, i.e. an SDF particle has to be transformed into a DE particle. The converse is true in the opposite direction. The other conversion is to do with schedulers. The schedulers in two interacting domains have to synchronize.

The above description gives an idea of how different models of computation are treated in Ptolemy. Different models of computation are kept pure. The main focus is placed on the interface between different models. The opposite approach in handling multiple models of computation is to compose them more tightly, mix their properties so that basically a new model results, which contains everything that its ingredients contain. The developers of Ptolemy criticize this brute-force approach because it results in what they call in [20] *emergent* behaviour. The designer expects that the resulting model of computation will have the properties of the models from which it was created. Instead, it often happens that the resulting model exhibits unexpected and undesired behaviour. It is also difficult to analyse the model.

On the other hand, it is questionable whether the approach in Ptolemy entirely preserves the properties of combined models. Some models are so different that it is really difficult to make them interact in a meaningful way.

In order to illustrate a system specification in Ptolemy this section focuses on combinations between FSM and other models of computation. The reason for choosing this combination is that it probably has the greatest chance of successfully representing mixed control/dataflow systems.

The combination of FSM and other models of computation is called *charts [85] in Ptolemy (pronounced star charts) where * denotes a wildcard indicating that it is possible to combine various models of computation with FSM. In *charts the concurrency semantics is decoupled from the FSM model. The concurrency semantics depends on the model that is combined with FSM. It is possible to have multiple concurrency semantics in a single system. In [85], the authors discuss combining

FSM with DF, DE and SR. In this section the FSM / DF combination is mostly discussed, in particular combining FSM with SDF (synchronous data flow).

Figure 2.23 shows a system that is composed of FSM and dataflow domains. There are five levels of hierarchy marked as (a), (b), (c), (d), (e). A hierarchical decomposition as in Fig. 2.23 is typical in Ptolemy. The hierarchy can be arbitrarily deep. Each hierarchical level belongs to a particular domain or model of computation. The model of computation determines how the blocks at that level communicate. For example the communication at level (a) is governed by SDF. Two or more models of computation cannot coexist in parallel at the same level of computation. In terms of what happens inside blocks, the black-box approach is taken. Blocks can be defined with different models of computation. It is only the communication between blocks that is governed by a single model of computation.

In *charts, an FSM can be used to describe a block. A state of an FSM can also be refined to another model. One of the problems with refining a state of an FSM in Ptolemy is that the refining system has to complete an iteration when the state is left. This is possible in SDF, for example, where a complete cycle can be defined as an iteration. A complete cycle returns buffers to their initial state. It would be more difficult to refine an FSM state with a Kahn process network, since a Kahn process network operates on infinite streams and cannot be divided into iterations.

A *homogenous* SDF actor consumes / produces only one token on each input/ output. A homogenous SDF actor fits naturally into the FSM model. It is more difficult to handle a nonhomogenous actor. For example, in Fig. 2.23 actor A consumes two tokens from input *a* and one token from input *b*, so it is a nonhomogenous actor (numbers in brackets indicate how many tokens are consumed or produced). It is refined to an FSM. At the FSM level, the two tokens from the input *a* are treated as events. It was decided [85] to order multiple events with the notation from the language Signal. *a* denotes the most recent event while *a$1* denotes the second most recent event. Suppose that the FSM is in the state α and the SDF actor fires. The FSM takes the transition from α to β if both *a* and *a$1* are present. The presence or absence of an event is explicitly encoded in the corresponding SDF token. The transition produces the output event *x* which appears on the output *x(1)* of the SDF actor A. The SDF actor D at level (d) is also refined to an FSM. The output of SDF actor B produces two tokens *y(2)*, but the FSM below produces only one event *y*. The interpretation is that *y* is present while *y$1* is always absent since it is not mentioned in the transition. The events *y* and *y$1* at level (e) are connected to *y(2)* at level (d).

At levels (d) and (e) an FSM is activated whenever the SDF block refined by the FSM fires. It is often useful not to activate an FSM when the higher level SDF block fires. This is especially the case when one or more states of the FSM are refined by SDF blocks. Therefore there are two types of firings associated with SDF/FSM combinations [85]. *Type A* firing is when an SDF block fires but the FSM that refines it is not activated, instead the system that refines the FSM is activated. *Type B* firing is when an SDF block fires and the FSM that refines it is activated, as well as the system that refines the FSM. For example suppose that the schedule of the SDF network at level (a) is {D,C,C,C,E} and the FSM at level B is in state β. The first two firings of C at level (a) will be of type A. The FSM at level (b) will simple ignore all events. It will just pass tokens down to level (c). The third firing of C will

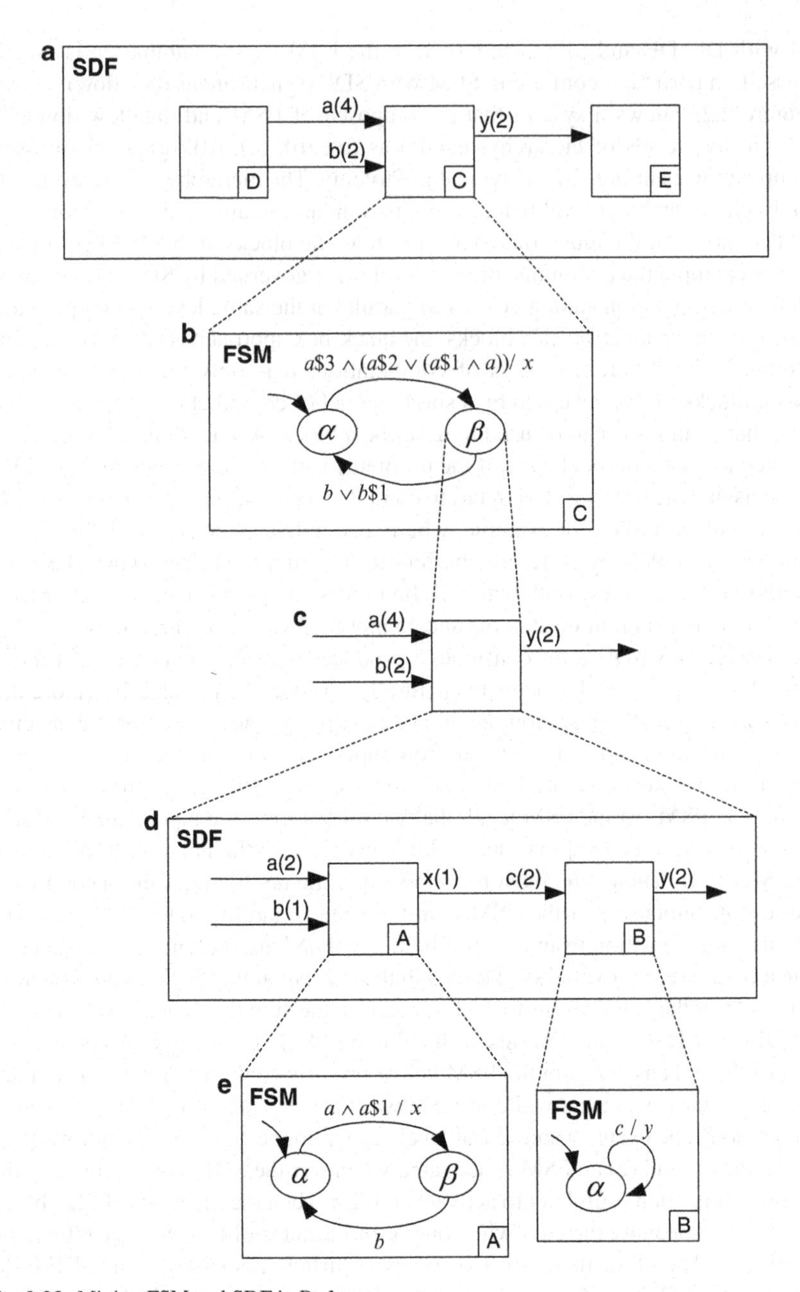

Fig. 2.23 Mixing FSM and SDF in Ptolemy

be of type B. This time the FSM will pass tokens to level (c), but it will also respond to them by producing the transition from β to α if the condition $b \vee b\$1$ is true.

Difficulties in combining FSM and SDF arise when states of an FSM are refined into SDF subsystems that do not consume the same number of tokens. If the FSM

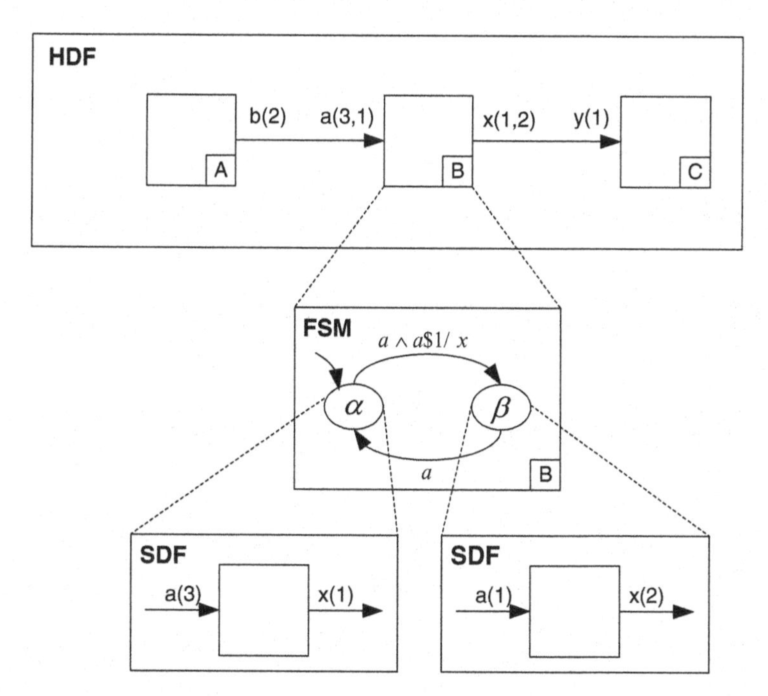

Fig. 2.24 HDF with FSM and SDF

refines an SDF block, the number of tokens consumed by the SDF block cannot be constant. Therefore, it seems that it is necessary to use dynamic dataflow. However, in that case, all advantages of SDF such as static scheduling would be lost. In [85], a new model called *heterochronous dataflow* (HDF) is proposed. In HDF, the numbers of tokens consumed by inputs and the numbers of tokens produced by outputs are not constant, but the list of possibilities is finite so static scheduling is still possible. An example of a system that uses HDF is shown in Fig. 2.24.

In Fig. 2.24, state α of the FSM is refined into an SDF subsystem that consumes three tokens and produces one token, while state β is refined into an SDF subsystem that consumes one token and produces two tokens. As a result of that, block A at the HDF level can either consume three tokens and produce one token or consume one token and produce two tokens. It is possible to do scheduling of an HDF system statically. Details can be found in [85].

In Ptolemy, the basic idea is to keep different models of computations separate and concentrate on their communication. Having each level of hierarchy defined by one model of computation should make the analysis of complex systems easier. Furthermore, hierarchical composition in Ptolemy encourages clean, modular design. Unfortunately, interaction between different models of computation is rarely straightforward. Models have to adjust to each other in order to be able to communicate, which often leads to restrictions in their expressive power.

Chapter 3
Specification in DFCharts

This chapter describes how embedded systems are specified in DFCharts. Section 3.1 presents an introduction to specification in DFCharts. Section 3.2 illustrates the application of DFCharts on a practical heterogeneous embedded system called frequency relay. Section 3.3 discusses languages and models related to DFCharts. Section 3.4 presents an extension of DFCharts called DDFCharts, which gives another dimension to DFCharts of being suitable to formally model class of distributed embedded systems.

3.1 Introduction to DFCharts

DFCharts targets heterogeneous embedded systems by mixing finite state machines (FSM) with synchronous dataflow graphs (SDFG). All FSMs in a DFCharts specification are driven by the same clock. On the other hand, each SDFG operates independently at its own speed. A DFCharts specification typically consists of multiple levels of hierarchy. The highest hierarchical level is referred to as "top level". FSMs and SDFGs that are placed at the top level are always active. Lower hierarchical levels are obtained by refining states of FSMs. When a state is entered, the objects that refine it are activated. They are terminated when the state is left.

3.1.1 Operators

The three most important operators in Argos (synchronous parallel, localization, and refinement) are also used in DFCharts to combine FSMs. The scope of the refinement operator is extended in DFCharts, and a state of an FSM can be refined not only by another FSM, but also by an SDFG. There are three ways in which a state of an FSM can be refined: only by FSMs; only by SDFGs; by FSMs and SDFGs at the same time,

I. Radojevic and Z. Salcic, *Embedded Systems Design Based on Formal Models of Computation*, DOI 10.1007/978-94-007-1594-3_3,

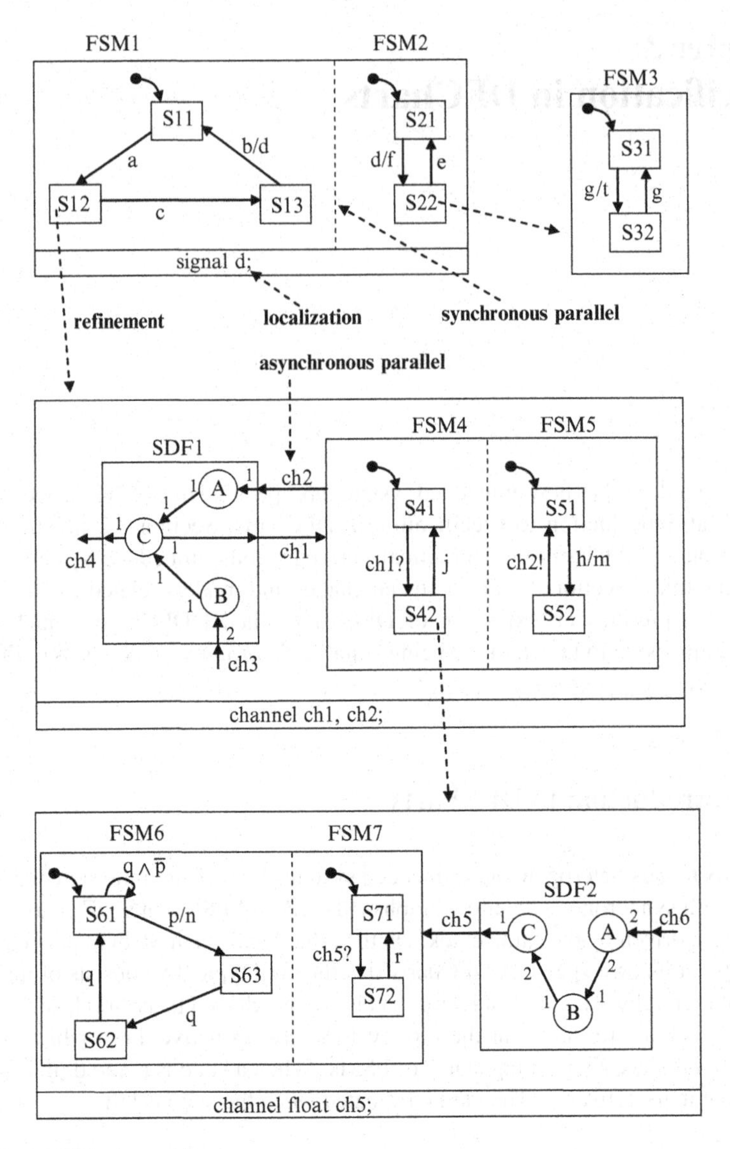

Fig. 3.1 A DFCharts specification

i.e. concurrently. The fourth operator of DFCharts, named asynchronous parallel, is used to connect an FSM and an SDFG on the same level of hierarchy. The use of the operators is illustrated in Fig. 3.1, which shows an example DFCharts specification. It consists of seven FSMs and two SDFGs, with three hierarchical levels.

The two top level FSMs, FSM1 and FSM2, execute concurrently as indicated by the synchronous parallel operator. In each tick of the clock, they read inputs and instantaneously produce outputs. Local signal d is used for synchronization. When

FSM1 is in S13 and input *b* is present, the transition to S11 is taken and *d* is emitted. If FSM2 is in state S21, *d* causes the transition from S21 to S22.

When state S22 of FSM2 is entered, FSM3 is started. When *e* is present, FSM2 leaves S22 and FSM3 is terminated in the same instant. Weak aborts are used in DFCharts as in Argos, which means that an FSM is allowed to produce outputs in the instant it is aborted. Thus, if both *e* and *g* are present, and FSM3 is S31, then *t* is emitted. Another type of refinement is illustrated in FSM1, where S12 state is refined by one SDFG and two FSMs.

An SDFG has two kinds of buffers: internal and interface buffers. Internal buffers are used for communication between actors within a graph. Interface buffers, which can be input or output, are used for communication between actors and FSMs or external environment. The execution of an SDFG consists of a series of iterations. Before each iteration, the SDFG has to receive inputs and send outputs produced during the previous iteration. As no outputs have been computed before the first iteration, the output buffers must contain initial tokens. Initial tokens can also be placed on internal buffers to prevent deadlock.

Input and output buffers are connected to the external environment and FSMs through channels. In Fig. 3.1, *ch3* and *ch4* connect SDF1 to the external environment, while *ch1* and *ch2* connect it to FSM4 and FSM5. The direction of each channel is indicated by an arrow. The communication between an SDFG and an FSM through a channel can occur only when both sides are ready for it. When the communication on a channel has taken place, both sides are notified by the event called rendezvous. It should be mentioned that in CSP [13], processes also have to meet in order to communicate. However, the procedure leading to rendezvous is slightly different. In CSP, both reads and writes are blocking. If the sender wants to send, but the receiver is not ready to receive, the sender will block. Similarly, if the receiver is ready to receive, but the sender is not ready to send, the receiver will block. In DFCharts, an SDFG cannot start the next iteration until communications on all of its channels have been completed. However, an FSM can abort waiting for rendezvous regardless of whether it is attempting to send or receive.

As mentioned previously, an SDFG is ready to communicate on all channels when it finishes an iteration. An FSM is ready to communicate on a single channel when it enters a rendezvous state. A state is called rendezvous state if it has an outgoing transition that is triggered by rendezvous. Rendezvous states cannot be refined. An FSM may spend many ticks in the rendezvous state waiting for rendezvous, but rendezvous itself is instantaneous – the communication on a channel happens in a single tick. In Fig. 3.1, the rendezvous states are S41, S52 and S71.

When FSM5 is in S52, it is ready to send on *ch2* as seen from the transition ch2!. The CSP notation is used where '?' is used to denote input, while '!' is used to denote output. When SDF1 is ready to receive on *ch2*, the rendezvous occurs triggering the transition ch2!, which leads to S51. In this example, ch2 is used purely for synchronization. SDF1 only receives a token that acknowledges that the rendezvous has been completed on the channel. There is no real data flowing through. More interesting examples with typed channels will be presented later in this section after the introduction of variables.

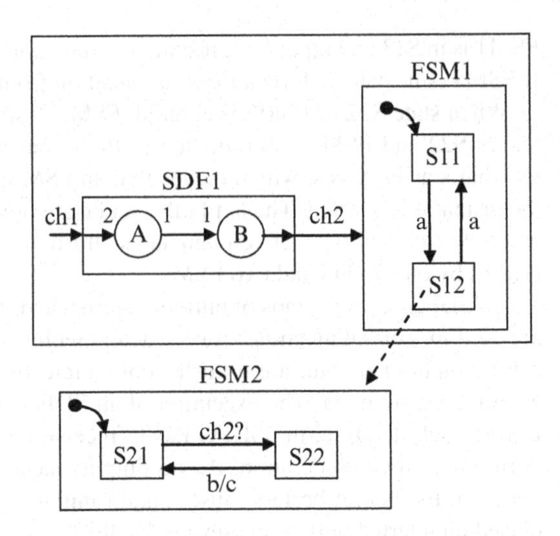

Fig. 3.2 Communication between an SDFG and a lower level FSM

For semantic purposes, we will assume that the communication between an SDFG and external environment also occurs when both sides are ready. However, in practical embedded applications, the external environment cannot wait. Data will be lost if the SDFG is not ready to receive it in its input buffers. Consequentially, there is a timing constraint that has to be satisfied. This issue will be dealt with in Chap. 7, which discusses implementation of DFCharts.

It should be noted that after each iteration of an SDFG, the rendezvous on each channel occurs only once. Suppose that FSM4 enters S42 in the instant when the rendezvous on *ch1* occurs. In the next tick of the FSM clock, j is present which causes FSM4 to go to S41 again. In the meantime SDF1 has not started a new iteration since the rendezvous on ch2 has not occurred yet. The rendezvous will not occur again on *ch1* until SDF1 completes a new iteration.

An FSM can be at a lower hierarchical level than an SDFG it communicates with. An example of that situation is given in Fig. 3.2. SDF1, placed at the top level communicates on *ch2* with FSM2 that refines state S12. This example also shows that waiting for rendezvous can be pre-empted by a higher level transition; in this case triggered by input signal *a*.

3.1.2 Transition Priorities

In Fig. 3.1, state S61 has an outgoing transition labelled $q \wedge \overline{p}$, where $\wedge$ denotes logical *and*, while the bar above *p* denotes logical *not*. Besides *and* and *not*, the logical connective *or* ($\vee$) can also be used in construction of Boolean expressions. The purpose of the *and* connective in this example is to make *p* and $q \wedge \overline{p}$ exclusive so that the execution of the FSM is deterministic. The transition triggered by *p* has the priority over the one triggered by $q \wedge \overline{p}$. Making triggers exclusive in this way can

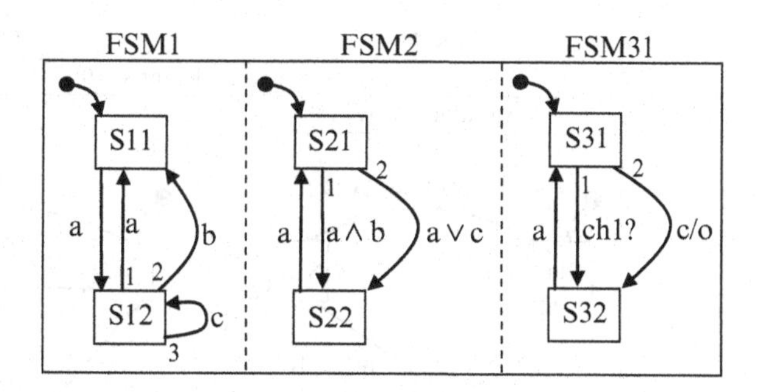

Fig. 3.3 Transition priorities

be difficult if a large number of signals are involved. In DFCharts, this problem can be handled by specifying transition priorities directly. This is illustrated in Fig. 3.3 by adding priority labels on transitions.

A rendezvous state can be left due to transitions that are triggered by signals. However, the transition that is triggered by rendezvous must always have the highest priority as shown in state S31. A rendezvous must happen if it is enabled.

3.1.3 Variables

The examples presented so far did not contain any data. Few practical applications can be specified by FSMs that only use pure signals. For this reason, variables can be employed in DFCharts. An FSM that uses variables becomes FSMD (FSM with datapath). Variables are also important for communication between FSMs and SDFGs. Under rare circumstances will an SDFG just synchronize with one or more FSMs, which was the case in Fig. 3.1. Usually, it will send and receive data.

DFCharts supports the use of variables, which can be shared or local. A local variable can only be used by a single FSM while shared variables can be used by multiple FSMs. However, among concurrently running FSMs, only a single FSM may write to a shared variable. The value of a local variable can change during a tick. On the other hand, a shared variable cannot have its value changed during a tick. If the writer updates a shared variable, the new value will be visible to the readers only in the next tick.

When variables are used, transitions can also contain conditions on variables and procedures that update variables. The general form of a transition is $t[c]/O,P$ where t is the transition trigger (either Boolean expression on signals or rendezvous), c is the condition on variables, O is the set of emitted output signals and P is the set of invoked procedures. Note that a transition cannot have a condition on variables if it is triggered by rendezvous.

Figure 3.4 shows a DFCharts specification with variables. Each variable has to be declared, and initialized if necessary. The declaration has to indicate the data type of the variable. In the Java environment implementing DFCharts, described in Chap. 6, only primitive data types float, integer and Boolean are currently supported for shared variables. However, the type of a local variable may be any data structure created in Java.

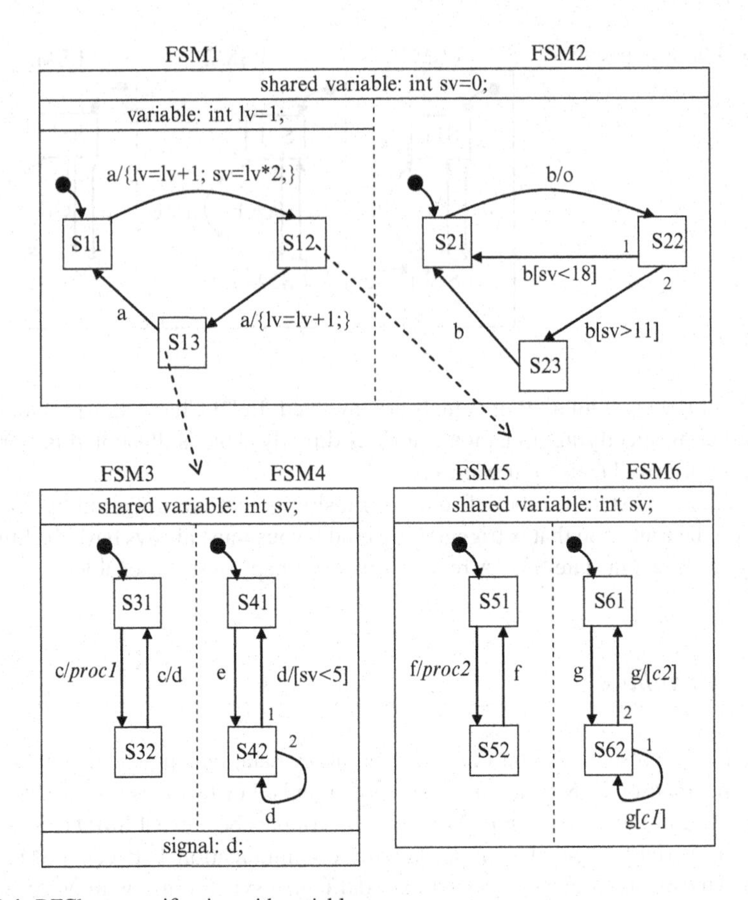

Fig. 3.4 DFCharts specification with variables

FSM1 contains two variables, local variable *lv* and shared variable *sv*. *lv* is only seen by FSM1 while *sv* can also be accessed by other FSMs that declare it. *sv* is written in the procedure that is called when FSM1 makes the transition from S11 to S12. In that procedure, *lv* is first incremented by one, and then *sv* gets the new value of *lv* multiplied by two. Since the value of a shared variable remains constant in a single tick, a procedure like {sv=sv+1; sv=sv*2;} would be pointless. The first statement would not have any effect on *sv*. The new value seen in the next tick will be determined by the last assignment, which is sv=sv*2 in this case. Smaller procedures, like the one just described, can be immediately shown in curly brackets. The names of the larger procedures such as *proc1* and *proc2* are given in italics and the contents are defined elsewhere.

FSM2 reads *sv* in conditions sv<18 and sv>11. Similarly to procedures, short conditions are immediately shown, while long ones like *c1* and *c2* in FSM6 are indicated in italics and defined elsewhere. Since FSM2 is just a reader of *sv*, all descendants from FSM2 would only be able to read *sv* as well. Descendants from FSM1 may also write *sv*, but in any single tick only one FSM may be the writer. FSM3 and FSM5, which write *sv* in procedures *proc1* and *proc2*, respectively, are both allowed to be writers since they cannot be active at the same time. On the other hand, FSM3 and FSM4 cannot both write. FSM4, like FSM6, can only read *sv*.

It should be pointed out that outgoing transitions of a hierarchical state that is refined by an FSM that writes a shared variable cannot contain procedures that write the same shared variable. This can be seen in Fig. 3.4. The transitions going out of S12 and S13 do not contain procedures that write *sv*, since FSM3 and FSM5 write *sv*. The purpose of this restriction is to prevent FSMs at different hierarchical levels to write the same shared variable simultaneously.

3.1.4 Data Transfer Between FSM and SDF

The condition that there may be only one active writer of a shared variable can easily be abandoned if resolution functions are introduced, as in VHDL for example. In fact, the behaviour of variables shared between FSMs is not an essential part of DFCharts semantics. A lot more important aspect, which is characteristic to DFCharts, is how variables are used to enable transfer of data between FSMs and SDFGs. This is where array variables become important. Figure 3.5 shows a DFCharts specification, which illustrates how variables are used in communication between FSMs and SDFGs.

When channels are used for data transfer, not just for mere synchronization as in Fig. 3.1, their data type must be declared. As in the case of shared variables, the DFCharts Java environment currently supports only the transfer of primitive data types across channels. Besides the data type, the declaration of a channel also indicates how many tokens pass through it when rendezvous occurs. In Fig. 3.5, *ch2* transfers two tokens, while all the others transfer only one.

When rendezvous occurs on *ch1*, FSM1 makes a transition from S11 to S12, but at the same time, the received integer is stored in shared variable *x*. FSM2 sends data from variable *y* on *ch2*, which is stored in the input buffer of actor A, which belongs to SDF1. The number next to the actor indicates that two tokens are stored. Thus, FSM2 has to send two integers. This in turn means that variable *y* has to be an array having two integers, as seen in its declaration. The elements of *y* are accessed using the usual array notation, which can be seen from the procedure called when FSM2 makes the transition of priority 1 from S1 to S2. Variables needed for transmission or reception of multiple tokens, such as *y*, must be arrays and not simply sets of elements where the order does not matter. The reason is that the order in which tokens are stored in SDF buffers is often significant.

FSM3 communicates with SDF2 and SDF3 using local variable *z*. In S31, it is ready to send the value contained in *z* on *ch6* to SDF2. In S32, it is ready to receive a value on *ch7* from SDF3 and store it in *z*. In the instant when rendezvous occurs, an implicit procedure occurs that copies the value from a variable to a channel or vice versa. Instead of ch7?z we could write ch7/{z=ch7_token} where *ch7_token* denotes the value that is received on *ch7*. Similarly, instead of ch6!z we could write ch6/ {ch6_token=z}. The transition from S32 to S31 also calls a procedure that multiplies *z* by two. It must be emphasized that the implicit rendezvous procedure is always executed first. The variable is first modified by rendezvous and then

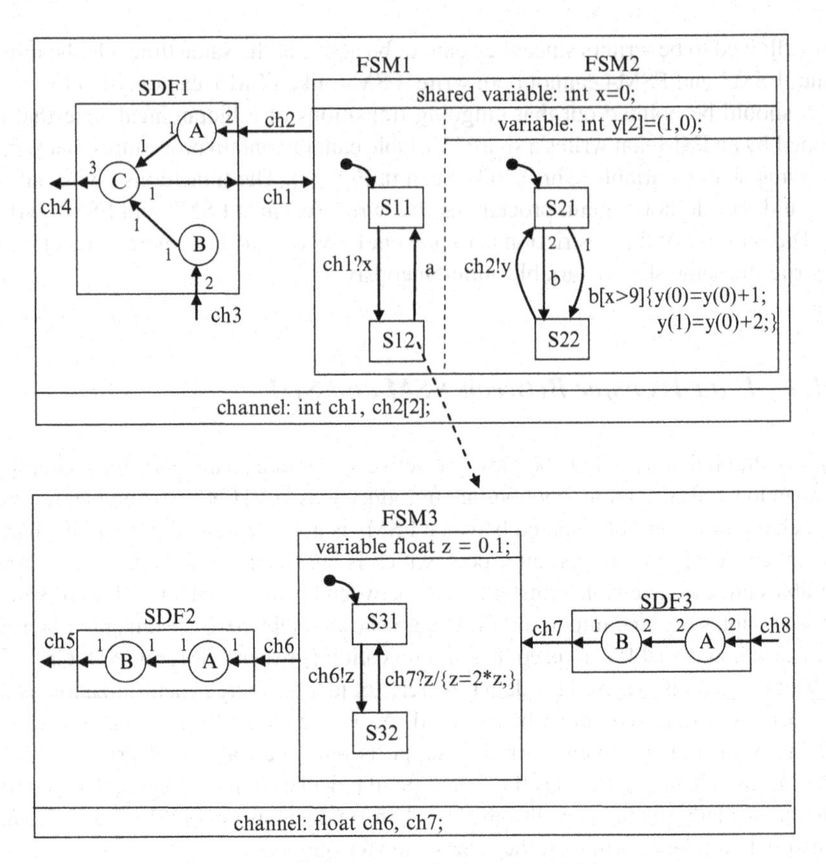

Fig. 3.5 Communication between FSMs and SDFGs through variables

multiplied by two. It cannot be the other way around. Otherwise, non-determinism could occur due to different execution orders.

When an FSM uses a shared variable for rendezvous, it must be the only active writer of the variable. This ensures that the value of the variable will not change while the FSM is in the rendezvous state.

3.2 Case Study: Frequency Relay

Power systems need protection from overloading. When a power system is overloaded some loads must be disconnected in order to prevent damage. A significant decrease in the frequency level of the main AC signal whose normal value is 50 (or 60) Hz indicates that the system may be dangerously overloaded. The same problem is also detected when the rate of change of the frequency of AC signal is too fast. The frequency relay is a system that measures the frequency and its rate of change in a power network.

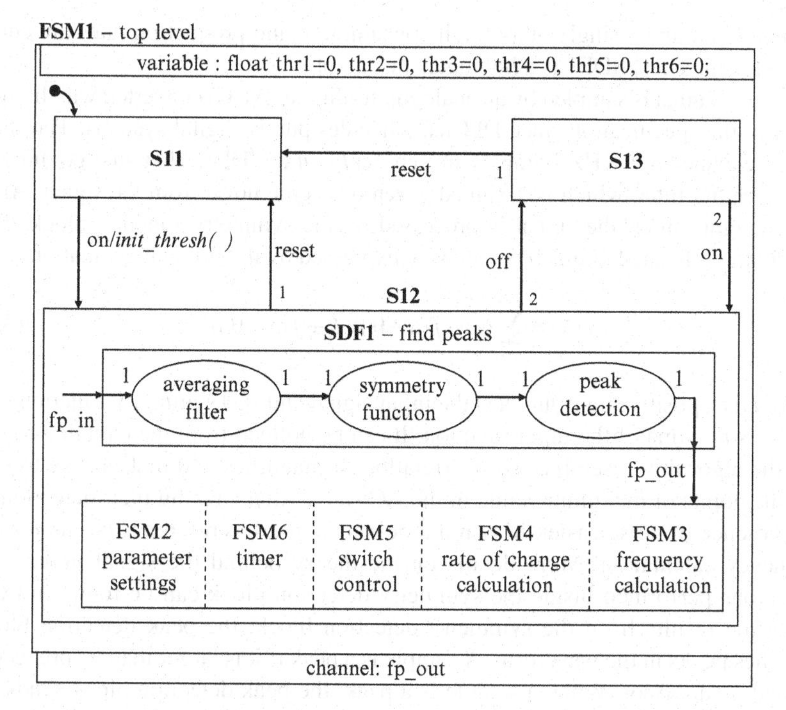

Fig. 3.6 Frequency relay – top level

Measurement results are compared against a set of thresholds that can be modified through a user interface. There are six thresholds in total, three for the frequency and three for the rate of change. If the current thresholds indicate that the frequency is too low or its rate of change too fast some loads are disconnected from the network by opening one or more switches (three in the presented case), as determined by a decision algorithm. Loads are gradually reconnected if the frequency and its rate of change improve. The specification consists of seven FSMs and one SDFG. In the following figures, we will show all signals and channels but only important variables.

Figure 3.6 shows the top level FSM. The main operation in the system occurs in S12 which is entered from the initial state S11 when *on* is present. In that transition, the thresholds are set to default values in *init_thresh() procedure*. S31, where nothing happens, is entered from S12 when *off* is present. S11 can be reached from the other two states with *reset.*

3.2.1 Peak Detection

S12 is refined by six FSMs and one SDFG. The purpose of SDF1 is to find the time between every two consecutive peaks in the AC waveform. With this information, the frequency can easily be determined. SDF1 consists of three blocks, each having a single input and a single output. All consumption and production rates are equal to one token. Thus, SDF1 is a homogeneous graph.

The AC signal is sampled by an analogue-to-digital (ADC) converter, which is not a part of the specification since DFCharts handles purely digital systems. The sampling frequency is 8 KHz. Samples are sent on *fp_in* to SDF1. They first go through the averaging filter, which is designed to remove some noise from the signal. After the averaging filter, the signal is processed by the symmetry function block. The algorithm performed in this block may be expressed by the following equation:

$$r(x) = \sum_{\theta=0}^{\pi/2} (L + f(x+\theta))(L + f(x-\theta)) \tag{3.1}$$

where L is a positive constant, $f(x)$ the input signal and $r(x)$ symmetry function that indicates maxima of the input function. It can be noticed from the equation above that the algorithm resembles autocorrelation. It magnifies and makes easy to isolate the points of maximum value in the AC signal, thus making their detection in the presence of noise easier. Maximum points are used as reference points for the frequency calculation. More details on the theory behind the signal processing operations performed inside the symmetry detection block can be found in [86]. Using the results from the symmetry detection block, the peak detection block identifies peaks in the waveform. A sample is a peak if it is larger than its predecessor and it successor. When a sample is a peak, the peak detection block sends on *ch1* the number of samples counted between the current peak and the previous one. Otherwise, zero is sent.

3.2.2 Frequency and Rate of Change Calculations

The data sent on *fp_out* is received by FSM3, shown in Fig. 3.7. In the instant when rendezvous takes place on ch1, FSM3 makes a transition from S31 to S32 while storing the number of samples in local variable *din*. In the next tick, the value of *din* is examined. If it is zero, the transition is taken back to S31, where a new rendezvous is awaited. Otherwise, the transition to S33 is taken, which calls procedure *af_calc()* (average frequency calculation). The instantaneous frequency is easily calculated by dividing the sampling frequency with the number of samples between the last two peaks. In order to smooth out any spikes and noise in the AC waveform, the frequency is averaged over four values. In the next tick, the average frequency value is compared with the three frequency thresholds (thr1, thr2 and thr3) in procedure *fs_calc()* (frequency status calculation). *Fs* (frequency status) is an integer that indicates the position of the average frequency with respect to the thresholds. It can take four values: 0 if ave_freq > thr1, 1 if thr1 > ave_freq > thr2, 2 if thr2 > ave_freq > thr3, and 3 if thr3 > ave_freq.

The rate of change of the frequency is handled by FSM4, shown in Fig. 3.7. It remains in the initial state S41 until FSM3 emits *start_roc*. When *start_roc* is present

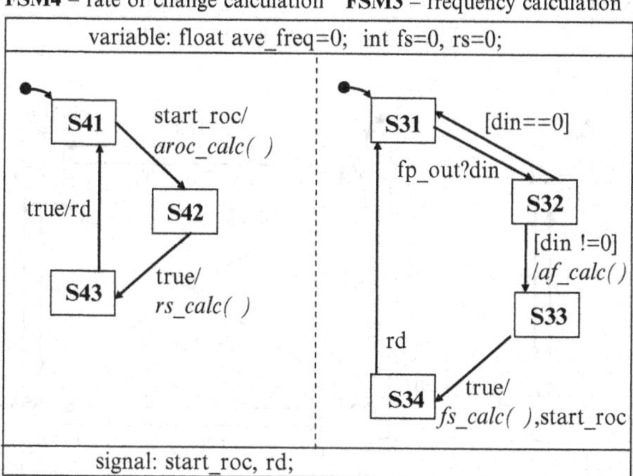

Fig. 3.7 Frequency and rate of change calculations

the transition from S41 to S42 is made calling *aroc_calc()* (average rate of change calculation) procedure. The instantaneous rate of change is calculated in the procedure from the current and previous values of the average frequency. The average rate of change is then calculated using four values. In the next tick, *rs_calc()* is called, which determines the value of *rs* (rate of change status) based on the average rate of change and the three related thresholds *th4*, *th5* and *th6*. The thresholds are used in exactly the same way as with *fs*. Thus, *rs* also takes one of the possible four values, 0, 1, 2 or 3. Finally, the transition from S43 to S41 emits *rd* (roc done) which brings FSM3 to the initial state and tells FSM5 to check the values of *fs* and *rs*.

3.2.3 Switch Control

FSM5, shown in Fig. 3.8, determines how many switches should be closed (turned on) using the values of *fs* and *rs*. In fact, each state of FSM5 directly corresponds to the state of the switches. In S51 all three switches are closed; in S52 two are closed; in S53 one is closed; all three are open in S54. The presence of *rd* signals that the values of *fs* and *rs* have been updated leading to a possible state change. The values of *fs* and *rs* are read in conditions *c1*, *c2*, and *c3*, which stand for $fs == 1 \vee rs == 1$, $fs == 2 \vee rs == 2$, and $fs == 3 \vee rs == 3$ respectively. Increasing values of *fs* and *rs* indicate increasing network load. FSM5 immediately has to respond by opening an appropriate number of switches. On the other hand, when the network condition begins to improve, which is marked by decreasing values of *fs* and *rs,* switches are reconnected one by one after a certain period of time. This is why there is a transition from S51 to S54, but not the other way around. To get from S54 to S51, FSM5 must go through S53 and S52.

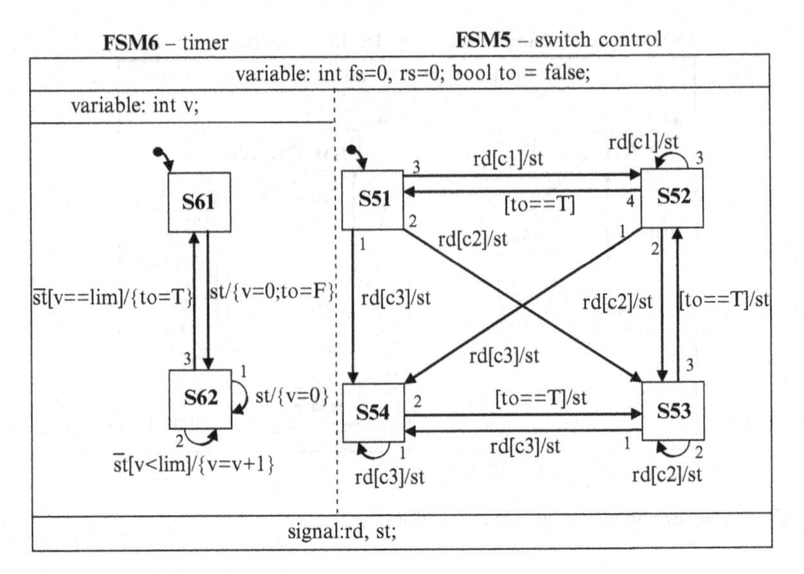

Fig. 3.8 Switch control and timer

The time needed for a switch to get reconnected represents a number of ticks, counted by FSM6 (Fig. 3.8). FSM5 and FSM6 communicate through the shared variable of type Boolean *to* (time out) and local signal *st* (start timer). All transitions in FSM5 start the timer except the one from S52 to S51. This transition does not start the timer since there are no more switches left to reconnect. When the timer is started, FSM6 makes the transition from S61 to S62 and sets the count variable *v* to zero. In each tick *v* is incremented until it reaches the limit. This process can be reset any time by another *st*. When the limit is reached FSM6 goes back to S61 with *to* becoming true. A transition can then be triggered in FSM5 with the condition to == T.

3.2.4 Threshold Modification

Threshold modification is handled by FSM2 and FSM7, shown in Figs. 3.9 and 3.10. The process of entering new thresholds begins when input *sth* (start thresholds) is present. FSM2 emits *inth* (input threshold) and enters state S22, which is refined by FSM7. Thresholds are received one by one in FSM7. In each transition *nt* (next threshold) is emitted, except for the transitions from S76 to S77, which emit *alld* (all done). Each threshold has two possible values that are selected by inputs *thresh0* and *thresh1*. It is also possible to leave a threshold unchanged using *skip*. The number of possible values for each threshold can be increased by using binary coding so that two inputs can cover four values. Moreover, the number of inputs could be increased. The selection is recorded with procedure *set_ft* for frequency thresholds and *set_rt* for rate of change thresholds.

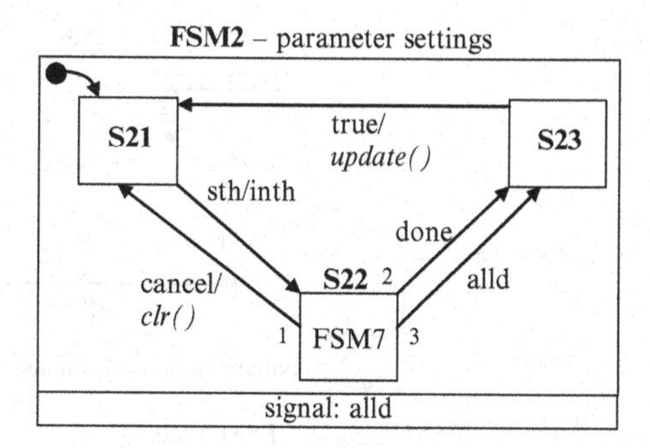

Fig. 3.9 Parameter settings

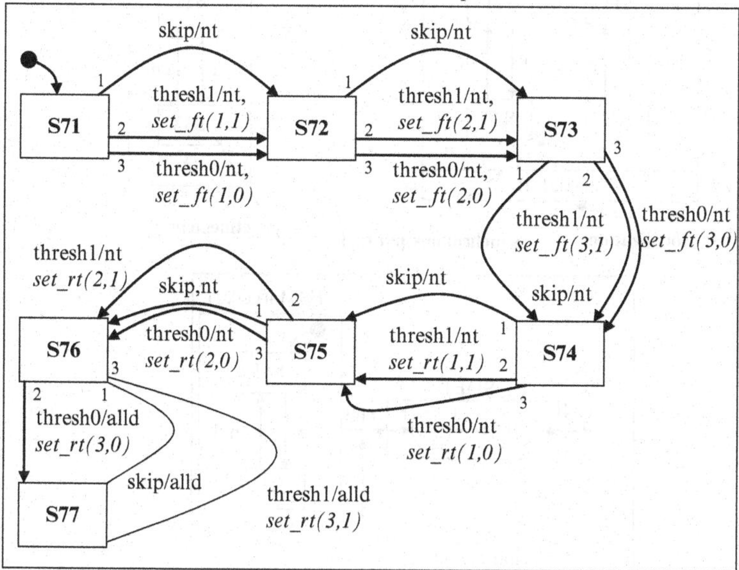

Fig. 3.10 Threshold reception

In FSM7, new threshold values are just recorded and do not take effect until procedure *update* in FSM2 is completed in the transition from S23 to S21. It is possible to pre-empt FSM7 either by *done* or *cancel*. When *done* occurs all new threshold values recorded before the pre-emption are written in *update*. In contrast, when *cancel* occurs all recorded values are cleared in procedure *clr*.

3.3 DDFCharts

Besides the mixture of control-dominated and data-dominated parts another important feature of complex embedded systems is distributed processing. Distributed systems are usually thought of as being comprised of physically distant processing elements,

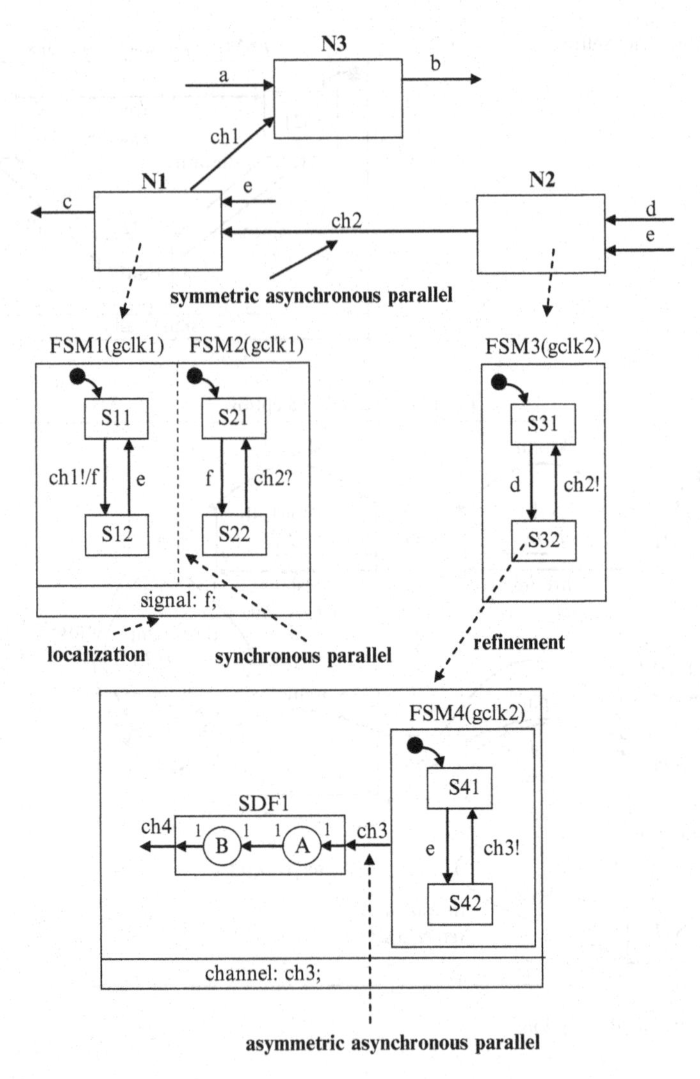

Fig. 3.11 A DDFCharts specification

but this type of structure can also exist on a single chip. A single DFCharts model is well suited for capturing the behaviour of a single localized system but it is more difficult to model a whole network. For the purpose of modelling distributed systems, an extension of DFCharts, called Distributed DFCharts (DDFCharts), is introduced. In a DDFCharts model, multiple DFCharts nodes are connected at the top level. In this definition, a DFCharts node is a single DFCharts model as described in previous sections.

Within each DFCharts node, four operators are used to connect FSMs and SDFGs as described in previous sections. However, an additional operator is needed in DDFCharts models to connect FSMs that belong to different nodes. This operator is based on asynchronous rendezvous communication. It is illustrated in Fig. 3.11 with a simple DDFCharts specification.

The DDFCharts specification in Fig. 3.11 is a network of three nodes, N1, N2 and N3. Only the contents of nodes N1 and N2 are shown. The semantics of synchronous parallel, asynchronous parallel, hiding and refinement operators is exactly the same as in DFCharts. Data handling with variables is also identical, although the simple example in Fig. 3.11 does not contain any. The asynchronous parallel operator is called asymmetric asynchronous parallel (AAP) operator in DDFCharts, to distinguish it from the symmetric asynchronous parallel (SAP) operator, which enables communication between DDFCharts nodes. The AAP operator connects an FSM and an SDFG within the same DDFCharts node, while the SAP operator connects two FSMs from two different DDFCharts nodes.

FSMs that are connected by the SAP operator are driven by different clocks. This is explicitly indicated in Fig. 3.11. For example, FSM2 from N1 is driven by gclk1, while FSM3 from N2 is driven by gclk2. The semantics of the SAP operator is very similar to the semantics of the AAP operator. Both sides must be ready for communication to occur. FSMs connected by the SAP operator are ready to communicate when they are in rendezvous states. The rendezvous states in FSM2 and FSM3 are S22 and S32 respectively. When these states are entered, the next ticks of gclk1 and gclk2 will occur simultaneously. Thus, transitions from S22 to S21 and S32 to S31 will be taken in the same instant.

3.4 Frequency Relay Extension

The frequency relay case study can be extended so that one or more parameters can be received remotely through a wireless CDMA link. As an example, the top level of one possible extension is shown in Fig. 3.12. In this version the constant *lim* used by FSM6 in Fig. 3.8 is now a parameter, which is received wirelessly using a CDMA receiver.

The system is specified in DDFCharts with two nodes at the top level as shown in Fig. 3.12. The top level of node N1 is shown in Fig. 3.13. It is almost unchanged from Sect. 3.2. The only difference is variable *lim* which is received through channel *tp* from node N2. Previously, it was a constant.

The purpose of node N2 is to receive the value of *lim* from a remote location and pass it on to node N1. It consists of two FSMs and two SDFGs as shown in Fig. 3.14.

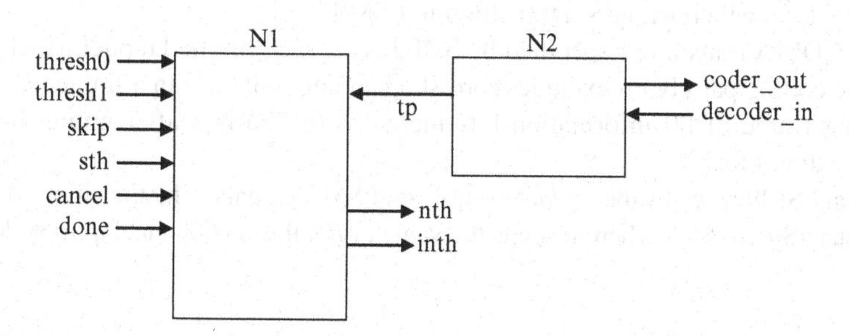

Fig. 3.12 Extended frequency relay – top level

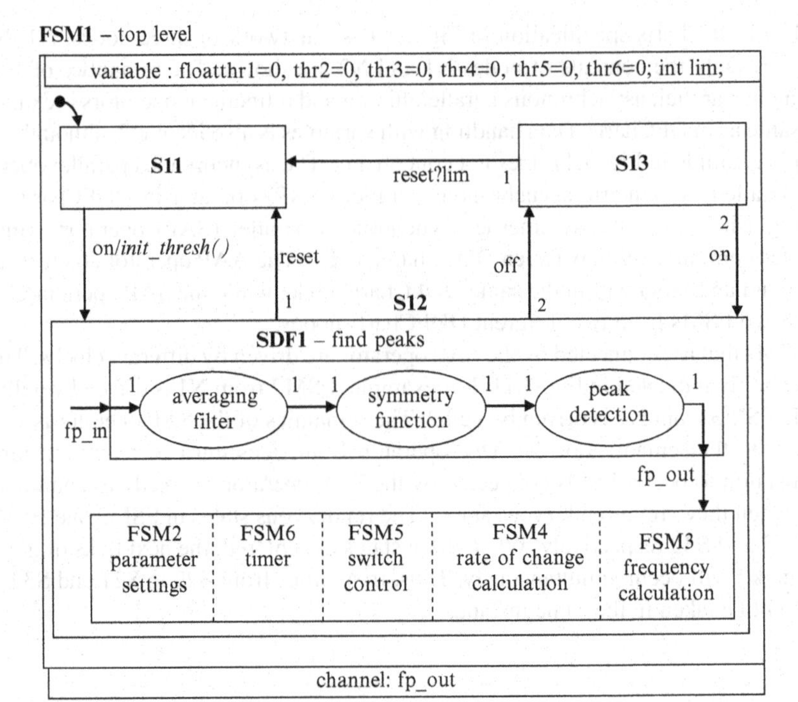

Fig. 3.13 Extended frequency relay – node N1

SDF3 represents a CDMA receiver which operates as described in [87]. Note that this is only a simplified model which contains the decoding part while omitting despreading and demodulation stages. In a single iteration, the receiver decodes an incoming packet of 384 bits to produce a frame of 80 information bits. In this case only 16 bits are needed to hold the value that is passed as an integer through channel *rx_out* to FSM8.

FSM8 receives the data from channel *rx_out* in state S81 and stores it in variable *lim* that can also be read by FSM9. In state S82, it sends the value of *lim* to SDF2 through channel *tx_in*. SDF2 represents a CDMA transmitter. The reason for transmitting back the value of *lim* to the remote location is simply to acknowledge its reception. When FSM8 makes the transition from state S82 to S81, and it also emits signal *update*, which triggers a transition in FSM9.

The CDMA transmitter specified by SDF2 is also a simplified model which contains the coding part, but it excludes spreading and modulation. In a single iteration, the input frame of 80 information bits increases to 576 bits after going through varios coding blocks.

When FSM9 receives the *update* event from FSM8, it makes the transition from its initial state S91 to S92, where it sends the value of *lim* through channel *tp* to node N1.

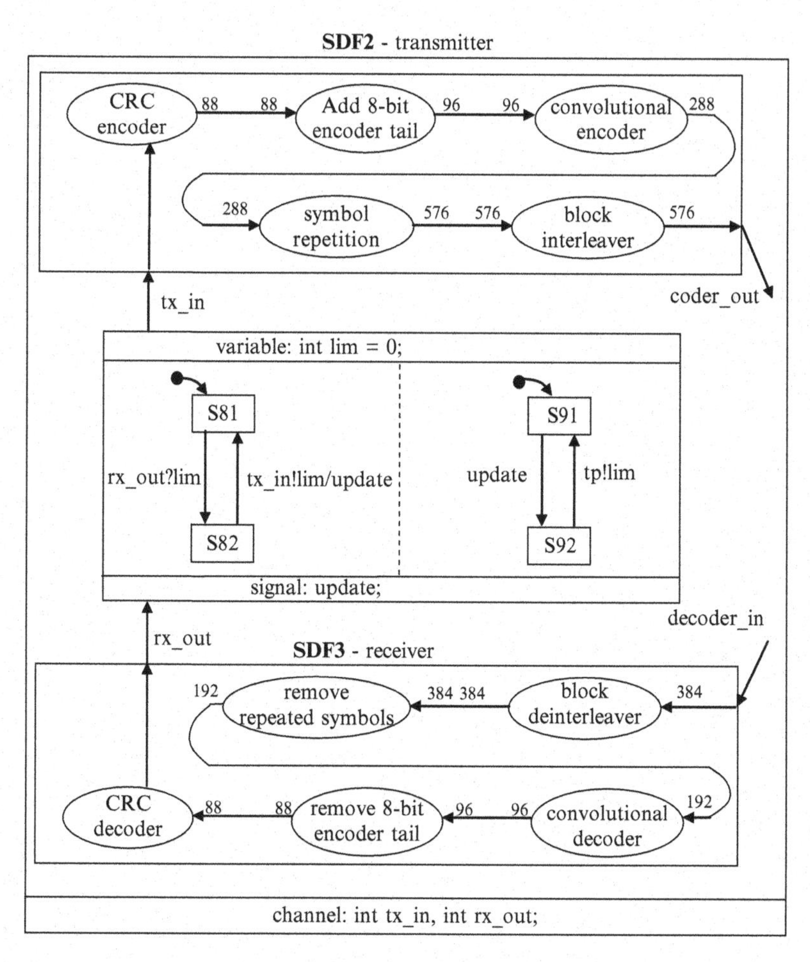

Fig. 3.14 Extended frequency relay – node N2

Chapter 4
Semantics of DFCharts

In this chapter we present the formal semantics of DFCharts. Section 4.1 discusses the automata semantics, introduced first in [88], where the behaviour of a complete specification is expressed as an FSM. This section is divided into seven sub-sections Section 4.1.1 introduces multiclock FSMs that are used as inputs to DFCharts operators. Sections 4.1.2–4.1.5 define the semantics of four DFCharts operators: synchronous parallel, asynchronous parallel, localization\hiding and refinement. DFCharts operators are defined in a similar style as Argos [24] operators. However, because of the use of multiclock FSMs, their operation is quite different. Section 4.1.6 defines with a simple language how the operators may be applied on multiclock FSMs. As seen in Chap. 3, an SDFG and an FSM are connected with the asynchronous parallel operator. According to the semantics of the asynchronous parallel operator in Sect. 4.1.3, it operates only on FSMs like all other operators. For this reason, an SDFG is represented as an FSM so that it can be combined with "real FSMs". This is the topic of Sect. 4.1.7. Since an SDFG proceeds at its own speeds, an FSM that represents an SDFG ("SDF FSM") is triggered by a clock that is different from the clock of real FSMs. Thus, when a "real FSM" and "SDF FSM" are combined, a multiclock FSM results. In the definitions of operators, which comprise DFCharts automata semantics, rendezvous is treated simply as an event that triggers a transition. What exactly happens on the channel is irrelevant. This is the topic of Sect. 4.2. It examines in detail the ordering of events on a channel within the Tagged Signal Model (TSM) framework. The analysis leads to understanding of how an array variable produces multiple SDF tokens and vice versa. Data transfer from SDF to FSM is the topic of Sect. 4.2.1, while Sect. 4.2.2.deals with data transfer from FSM to SDF. The TSM semantics of DFCharts was previously described in [89]. Finally, Sect. 4.3 examines the effect of clock speeds on DFCharts behaviour.

I. Radojevic and Z. Salcic, *Embedded Systems Design Based on Formal Models of Computation*, DOI 10.1007/978-94-007-1594-3_4,

4.1 Automata Semantics

In DFCharts automata semantics, the behaviour of a specification is determined by its equivalent, flat FSM, as in Argos. The equivalent FSM is obtained by combining FSMs and SDFGs using four operators: synchronous parallel (||), asynchronous parallel (\\), hiding (\) and refinement (↓ marks the state refinement by FSMs with or without SDFGs while ⇓ marks the state refinement only by an SDFG). All four DFCharts operators are associative. The synchronous parallel operator is also commutative. The equivalent FSM is constructed bottom-up, by starting from the lowest hierarchical levels and moving upwards to the top level. Inside a state, all FSMs are firstly combined by the synchronous parallel operator with the hiding operator applied if there are any local signals. Then, the equivalent FSM is combined with SDFGs by the asynchronous parallel operator to produce another single flat FSM. The resulting FSM is then combined with the higher-level FSM by the refinement operator.

Of course, the process described above would not be possible without representing the operation of an SDFG as an FSM, which will be discussed in Sect. 4.1.7. While "real FSMs" are all driven by the same clock, which we will call *gclk* (global clock), "SDF FSMs" are driven by their own individual clocks. Each clock in DFCharts semantics is a sequence of ticks. As in synchronous languages, each tick denotes a reaction. Clocks of "SDF FSMs" synchronize with *gclk* only when a rendezvous occurs. It is important to mention that *gclk* can start and stop SDF clocks. The relation between *gclk* and SDF clocks is illustrated with the example in Fig. 4.1. Ticks occur simultaneously only at rendezvous points when two clocks are synchronized, as indicated by dotted lines. Ticks cannot occur simultaneously by chance. This assumption needs to be made to ensure deterministic behaviour. As indicated in Fig. 4.1, it is possible that ticks of two SDFclocks occur simultaneously, but this is only the result of their simultaneous synchronization with *gclk*. SDF clocks cannot synchronize directly. It should be noted that SDF is not the only model that can be abstractly represented as an FSM and connected with FSMs through channels. Other dataflow models, which have clearly defined iteration, can easily be incorporated.

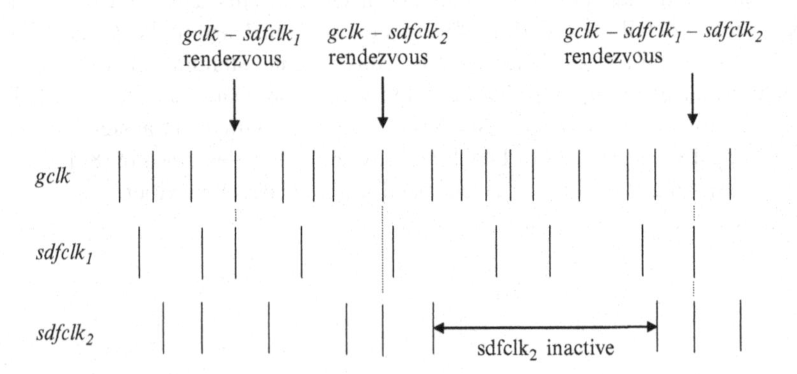

Fig. 4.1 Clocks in DFCharts

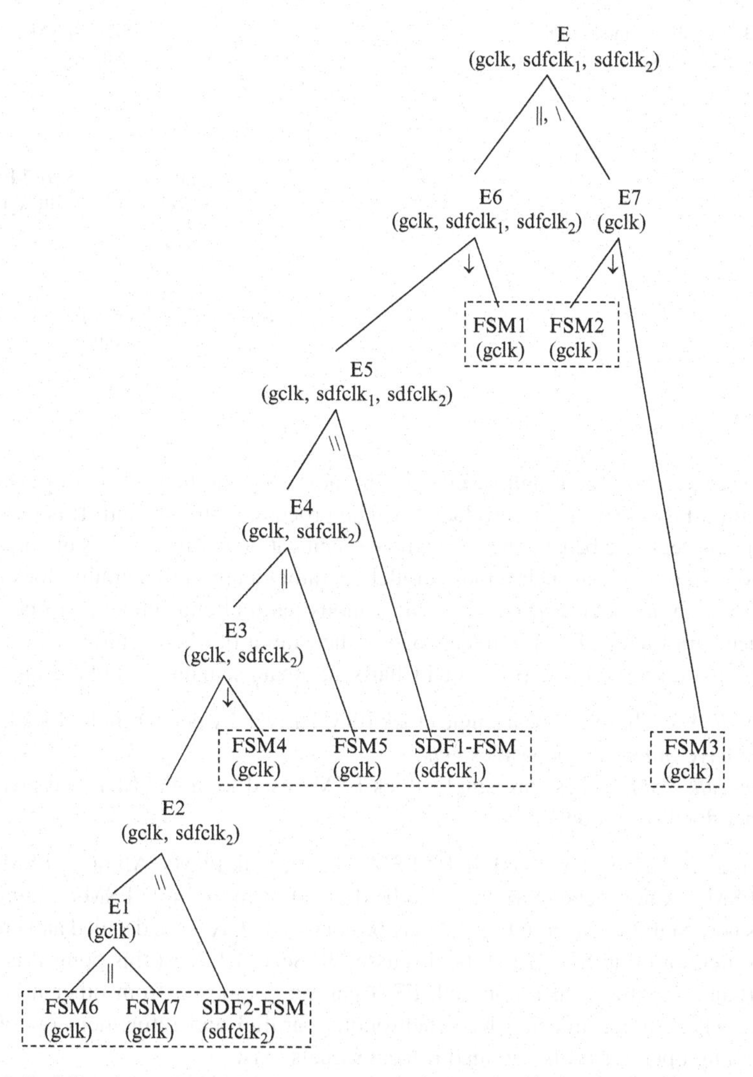

Fig. 4.2 Building the equivalent FSM for the example in Fig. 3.1

The steps for building the equivalent FSM E for the example in Fig. 3.1, which is reproduced here, are shown in Fig. 4.2. E1 to E7 are intermediate equivalent FSMs created in the process. Each step indicates the operator that is applied. Also, the clock(s) for each FSM are given in brackets. FSMs that have more than one clock are called *multiclock FSMs*. FSMs that refine the same state are placed in a rectangle. We explain a few steps in the beginning. At the lowest level of hierarchy are FSM6, FSM7 and SDF2. FSM6 and FSM7 are firstly combined with the synchronous parallel operator producing the equivalent FSM E1. E1 and SDF2-FSM are then combined with the asynchronous parallel operator producing the equivalent FSM E2. Since the clock of E1 is gclk and the clock of SDF2-FSM is $sdfclk_2$,

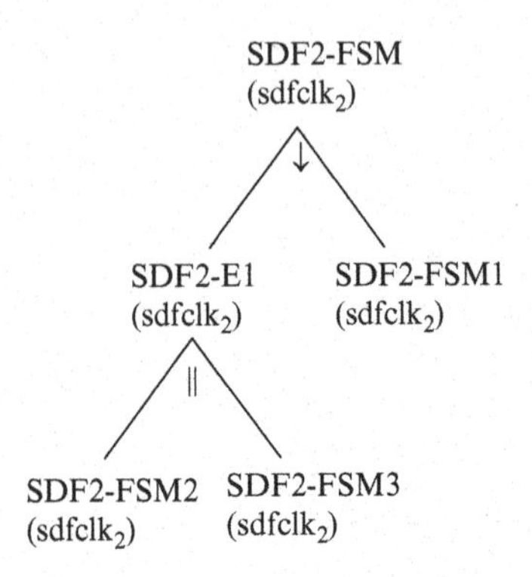

Fig. 4.3 Possible composition of a non-gclk FSM

E2 must have both gclk and $sdfclk_2$. In the next step the hierarchical operator is applied on FSM4 and E2 to produce E3. The process continues until E is reached, which represents the behaviour of the whole specification in Fig. 3.1. It is important to observe that only the asynchronous parallel operator can produce a multi-clock FSM from two single-clock FSMs. The synchronous parallel and refinement operators produce a multiclock FSM if at least one of their input FSMs is multiclock.

We divide FSMs that appear in DFCharts automata semantics in two groups:

- *gclk FSMs*: These can be a single clock FSMs driven by *gclk* or multiclock FSMs that have other clocks besides *gclk*.
- *non-gclk FSMs*: These are single clock FSMs used to model SDF and possibly other dataflow models.

In Fig. 4.2, SDF2-FSM and SDF1-FSM are non-gclk FSMs. All other FSMs are gclk FSMs. A non-gclk FSM may also be derived from multiple FSMs. Figure 4.3 shows how SDF2-FSM from Fig. 4.2 may be constructed. A more detailed and precise representation of an SDFG will be discussed in Sect. 4.1.7. At this point it is only important to realize that a non-gclk FSM can be composed from other non-gclk FSMs, which are connected by synchronous parallel and refinement operators. The hiding operator is also applied if local signals exist.

Since multiclock FSM E is too large to be drawn on a single page we present a simple multiclock FSM in Fig. 4.4. Due to space constraints *gclk* is written as *k*. The additional clock in the FSM is k1. This notation will be followed in other figures showing multiclock FSMs – *gclk* will be labelled as *k*, while other clocks will be labelled k1, k2,... At the beginning of each transition label, it is indicated which clock drives it. Synchronization of clocks is indicated by using brackets in the clock label. All clocks that are synchronized with *k* are placed in brackets. In a multiclock FSM, inputs are tied to specific clocks. At each tick of *k*, input *a* is read. On the other hand, input *b* is tied to clock *k1*. When *k* and *k1* are synchronized, in which case the clock label is <k(k1)>, *a* and *b* are read at the same time. A dot between inputs in monomial means logical AND, while a bar over an input means logical NOT.

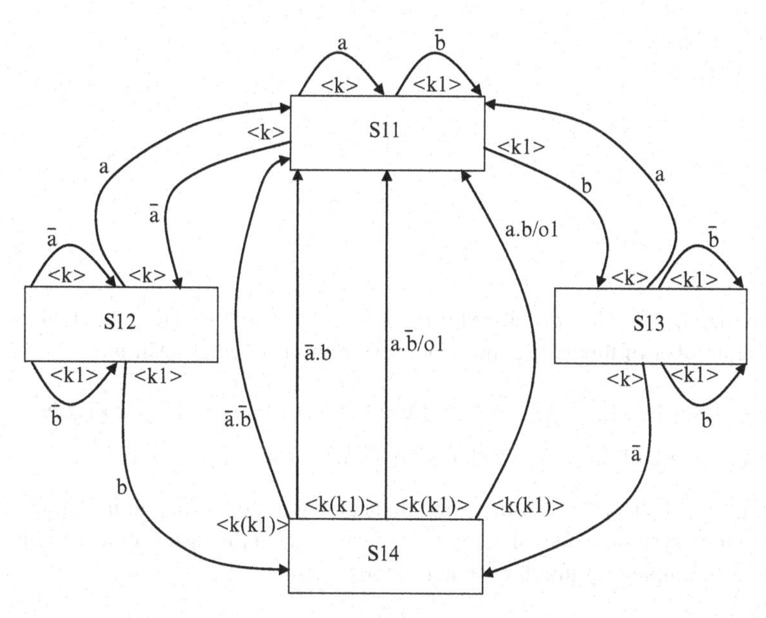

Fig. 4.4 A multiclock FSM

An alternative to having multiclock FSMs is to treat clocks as boolean inputs. A base clock is introduced which is faster than any other clock. When a tick of the base clock occurs, a transition will be taken if both its clock and input monomial are true. Since there is effectively a single clock triggering all transitions the semantics is likely to be simpler. This approach is used in Multiclock Esterel [58].

While it is desirable for the semantics to be as simple as possible, it should also reflect the true behaviour of the system as closely as possible. The base clock is fictitious. It does not show up in implementation. Moreover, it may be difficult to think of clocks as inputs. In pure hardware implementation, which Multiclock Esterel targets, clocks do appear as input signal lines in the synthesized circuit. This is not the case in software implementation, where a clock tick could represent the execution of several instructions.

In DFCharts, treating clocks as inputs is further complicated by the synchronization of clocks. When two FSMs are ready to communicate from rendezvous states, the next ticks of their clocks must occur simultaneously. If the clocks are viewed as inputs, they must be both true in the next tick of the fictitious base clock. Forcing clocks to have the same status means that they must be considered as a special kind of inputs. As a result, the definitions of the operators for composition would have to contain conditions attached to clocks. In the end, the single clock semantics may not be significantly simpler than the semantics based on multiclock FSMs that is presented in the following sections.

Reading asynchronous clocks like any other external inputs could result in many unnecessary transitions, thus making analysis more difficult. As seen in Fig. 4.4, there are four transitions going out of every state where clocks k and k1 are not

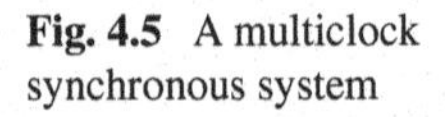
Fig. 4.5 A multiclock synchronous system

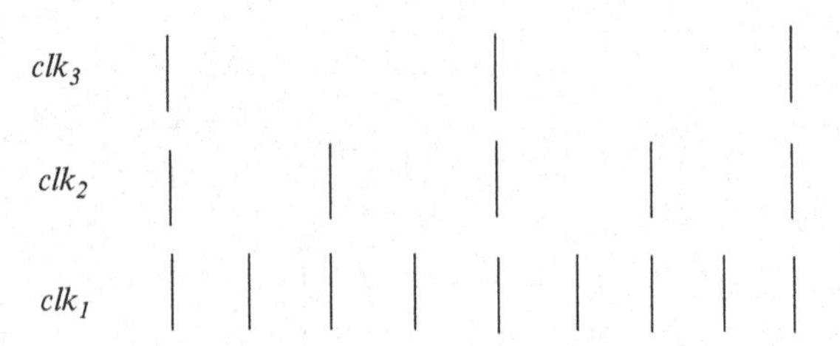

synchronized: $< k > a, < k > \overline{a}, < k1 > b, < k2 > \overline{b}$. If k and $k1$ are treated as inputs, the total number of inputs becomes four resulting in sixteen transitions:

$$k.k1.a.b, k.k1.a.\overline{b}, k.k1.\overline{a}.b, k.k1.\overline{a}.\overline{b}, k.\overline{k1}.a.b, k.\overline{k1}.a.\overline{b}, k.\overline{k1}.\overline{a}.b, k.\overline{k1}.\overline{a}.\overline{b},$$
$$\overline{k}.k1.a.b, \overline{k}.k1.a.\overline{b}, \overline{k}.k1.\overline{a}.b, \overline{k}.k1.\overline{a}.\overline{b}, \overline{k}.\overline{k1}.a.b, \overline{k}.\overline{k1}.a.\overline{b}, \overline{k}.\overline{k1}.\overline{a}.b, \overline{k}.\overline{k1}.\overline{a}.\overline{b}$$

Including clocks in transition monomials is much more efficient in a multiclock synchronous system, where all clocks are derived from the fastest clock. Figure 4.5 shows an example of multiclock synchronous system.

4.1.1 FSM with Variables

Definition 4.1. Finite state machine with variables

$$A = (CLK, Q, q_0, I, R, O, V, C, T, RQ, Proc)$$

CLK is the set of clocks that drive FSM transitions. For gclk FSMs, $gclk \in CLK$. For non-gclk FSMs, $gclk \notin CLK$ and $|CLK| = 1$. Q is set of states where q_0 is the initial state. I is the set of input signals. Each input signal can either be present (true) or absent (false). R is the set of channel status signals (CS signals). It is composed of external CS signals R_E and internal CS signals R_I, $R = R_E \cup R_I$. In DFCharts automata semantics, CS signals are tested in the same way as input signals. When a clock tick occurs, a CS signal is true if a rendezvous is happening on the corresponding channel and false otherwise. O is the set of output signals. Each output signal can either be emitted (true) or not emitted (false). V is the set of variables. C is the set of conditions on variables. Like input signals, CS signals and output signals, conditions are Boolean; they can be true or false. T is the set of transitions. RQ is the rendezvous mapping function. T and RQ will be defined below. *Proc* is the set of procedures on variables.

I, O, V, C, Proc and R_E are partitioned (divided into disjoint subsets) by clocks. On the other hand, internal CS signals R_I are shared among clocks as they are used for synchronization. Internal CS signals disappear when the asynchronous parallel operator is applied (Sect. 4.1.3).

It was mentioned in Sect. 3.1.1 that each rendezvous state is used for communication on a single channel. However, when an equivalent FSM is obtained by combining two or more FSMs, a state may be linked to multiple channels. *RQ*, defined below, is a function that maps CS signals to states. Each state is mapped to zero or more CS signals.

Definition 4.2. Rendezvous mapping function

$$RQ : q \rightarrow 2^{R}$$

Definition 4.3. FSM transitions

$$T \subseteq CLK \times 2^{CLK} \times Q \times B(I') \times B(C') \times B(R') \times 2^{O} \times Proc \times Q$$

A transition is driven by a clock *clk* taken from the set *CLK*. In addition, in the case of gclk multiclock FSMs, there could be a set of clocks taken from 2^{CLK} (power set of CLK) that have to synchronize with *clk* when the transition occurs. This is applicable only when $clk = gclk$. $B(I')$ where $I' \subseteq I$ is the set of Boolean formulas on input signals that are bound to *clk*. Each formula is a complete monomial, or in other words a conjunction that has to contain every input signal in I'. $B(C')$ where $C' \subseteq C$ is the set of Boolean formulas on variable conditions that are bound to *clk*. Each formula is a complete monomial. $B(R')$ where $R' \subseteq R$ is the set Boolean formulas (again complete monomials) on CS signals that are linked to the source state of the transition and bound to *clk*.

We can write a transition as a tuple (clk, q, i, o, p, q'). $i = m.c.r$, where *m*, *c* and *r* are monomials over input signals, conditions on variables and CS signals, respectively. The dot between monomials denotes the AND operation exactly in the same way as inside monomials. *o* and *p* denote output signals and procedures.

Complete monomials are required for the analysis of determinism and reactivity. DFCharts examples given in Sect. 3.1 never contained complete monomials. The requirement to specify complete monomials would significantly increase specification time, especially when there are many conditions on variables. Fortunately, it is not necessary since complete monomials can always be produced from incomplete monomials by adding transitions. We illustrate this with two examples in Figs. 4.6 and 4.7. The first specification (Fig. 4.6a) has one input signal and two conditions on variables, the second (Fig. 4.7a) has two input signals and one channel. In both cases, all transitions are driven by *gclk*.. In Fig. 4.6b, b and c stand for conditions

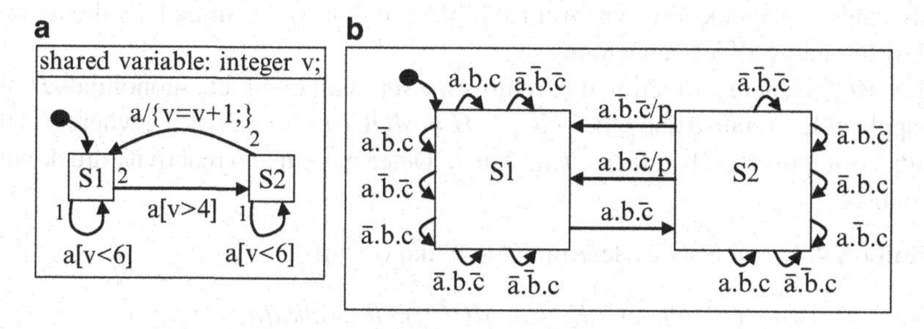

Fig. 4.6 A specification with one input signal and two conditions

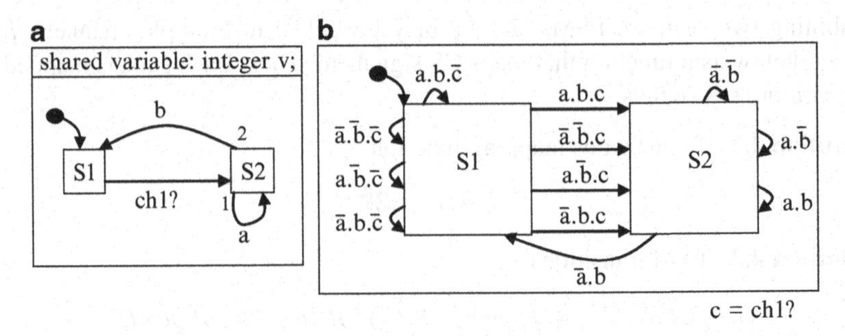

Fig. 4.7 A specification with two input signals and one channel

[v>4] and [v<6], while p stands for procedure {v=v+1}. In Fig. 4.7b, CS signal c corresponds to ch1?.

It can be observed in Fig. 4.6b that data abstraction is used when conditions on variables are represented, since values of variables are not visible. For each condition, it only matters whether it is true or false. Explicit representation of variables would make any analysis, including causality related properties of determinism and reactivity, impossible, since each variable can take an infinite number of values. A similar approach for data abstraction is used in Esterel Studio [18]. There are many other approaches for ensuring that the state space is finite including symbolic bisimulation [90] in CCS and region graphs in timed automata [91]. In each state, all input signals and all conditions are checked. However, this is not the case with CS signals. A CS signal is only checked in the states it is associated with. The variability in the number of outgoing transitions is seen in Fig. 4.7b, where S1 has 8 outgoing transitions and S2 has four outgoing transitions. For definitions of reactivity and determinism, the variability in inputs from state to state does not pose a problem. However, multiple clock domains are more difficult to deal with. A clock domain is simply a set of FSMs that are driven by the same clock.

An obvious approach in handling reactivity and determinism for a multi-clock FSM is to check them separately for each clock domain. Then we can state that an FSM is deterministic and reactive if determinism and reactivity is satisfied in each clock domain. In this discussion we are not concerned with the relation between clocks. We only assume that ticks of two different clocks never occur simultaneously unless the clocks are synchronized. In Sect. 4.3, we examine how the relative speeds of clocks affect behaviour.

Let *MQC* denote a function that returns the set of all complete monomials *H* that are applicable to state *q* and clock *clk* i.e. $H = MQC(q, clk)$. $|H| = 2^n$, where *n* is the number of inputs read by clock *clk* in state *q*. Determinism and reactivity are defined as follows.

Definition 4.4. An FSM is deterministic if and only if

$$\forall q \in Q, \forall clk \in CLK, \forall i \in MQC(q, clk), (clk, q, i, o_1, p_1, q_1')$$
$$\in T \wedge (clk, q, i, o_2, p_2, q_2') \in T \Rightarrow (o_1 = o_2) \wedge (p_1 = p_2) \wedge (q_1' = q_2')$$

In every state, for every clock domain, no two transitions may have the same input combination in order for the FSM to be deterministic.

Definition 4.5. An FSM is reactive if and only if

$$\forall q \in Q, \forall clk \in CLK, \forall i \in MQC(q, clk), \exists (clk, q, i, o, p, q') \in T$$

In every state, for every clock domain, there has to be at least one transition for each input combination in order for the FSM to be reactive.

Before giving the definitions of the operators, we present synchronous product and asynchronous product (or interleaving), two well known methods used for composing single clock FSMs. Synchronous parallel, asynchronous parallel and refinement operators use both synchronous product and interleaving.

Synchronous product is shown in Fig. 4.8. The outgoing transitions from states A and B are synchronously combined to produce the outgoing transitions from state AB. This means that each transition monomial in state AB is obtained by ANDing monomials from states A and B. In Fig. 4.8, all transitions are driven by the same clock, but synchronous product may also be applied when two different clocks are synchronized as shown in Fig. 4.9. Asynchronous product is simply interleaving of transitions driven by two different clocks when they are not synchronized. It is shown in Fig. 4.10. The transitions from states A and B are unchanged in state AB, they are simply copied. It is important to emphasize that transitions driven by *k* and *k1* are never enabled at the same time. Thus, the asynchronous product is deterministic due to the explicit use of clocks.

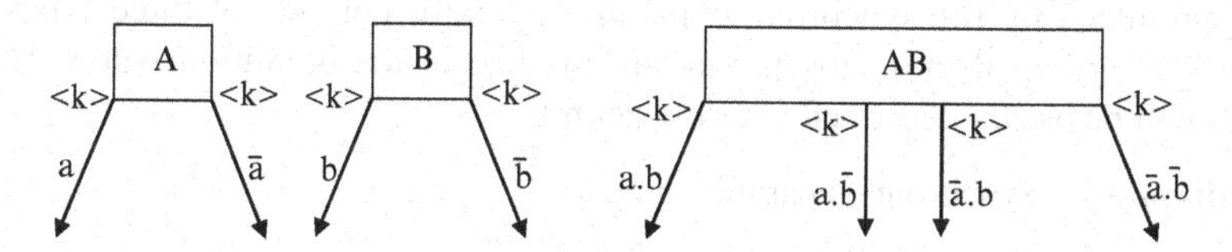

Fig. 4.8 Synchronous product with a single clock

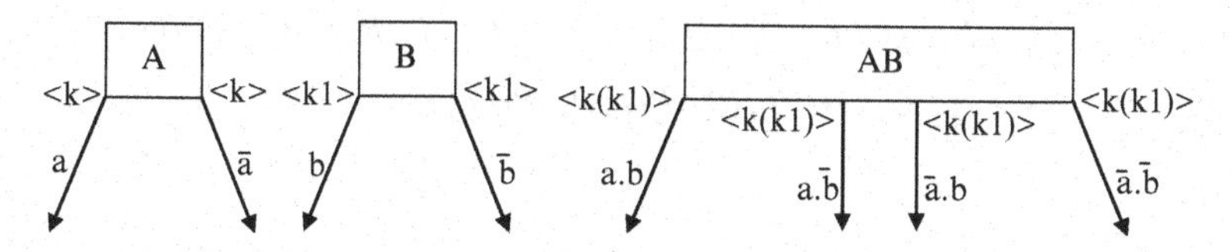

Fig. 4.9 Synchronous product with two synchronized clocks

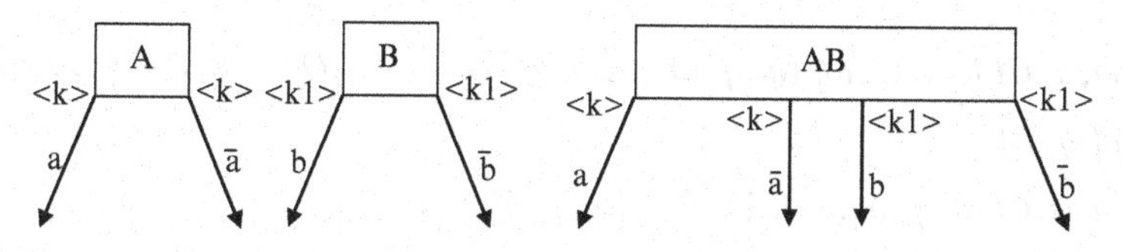

Fig. 4.10 Asynchronous product

When the asynchronous parallel and hiding operators are applied, internal signals disappear. The removal of signals is done with the hiding function defined below.

Definition 4.6. Hiding Function

$$HF : M \times 2^S \rightarrow M$$

S is the set of all inputs. M is the set of all possible monomials that can be formed from inputs in set S. $HF(m,I) = m'$ where m' is the monomial that is obtained when signals in set I are removed from monomial m.

We also define additional notation for monomials before the definitions of the operators. For a monomial m, m^a is the set of all signals that appear in m, m^+ is the set of signals that appear as positive elements in m while m^- the set of signals that appear as negative elements in m. For example, for $m = a.\overline{b}.\overline{c}$, $m^a = \{a,b,c\}$, $m^+ = \{a\}$ and $m^- = \{b,c\}$.

4.1.2 Synchronous Parallel Operator

The synchronous parallel operator can connect two gclk FSMs or two non-gclk FSMs driven by the same clock. FSMs connected by the synchronous parallel operator must not have any common CS signals. When the definition below is applied on two gclk FSMs, *mclk* stands for *gclk*. When two non-glck FSMs are involved, *mclk* represents the single clock of the two FSMs. The transition set of the FSM produced by the synchronous parallel operator consists of three types of transitions, grouped into subsets that are marked in the definition below. Each subset will be briefly described after the definition.

Definition 4.7. Synchronous parallel operator

$$\begin{aligned}
&(CLK_1, Q_1, q_{01}, I_1, R_1, O_1, V_1, C_1, T_1, RQ_1, Proc_1) \parallel \\
&(CLK_2, Q_2, q_{02}, I_2, R_2, O_2, V_2, C_2, T_2, RQ_2, Proc_2) \\
&= (CLK_1 \cup CLK_2, Q_1 \times Q_2, (q_{01}, q_{02}), I_1 \cup I_2, R_1 \cup R_2, O_1 \cup O_2, \\
&V_1 \cup V_2, C_1 \cup C_2\, T, RQ, Proc_1 \cup Proc_2)
\end{aligned}$$

where

$$RQ_1(q_1) = S_1 \wedge RQ_2(q_2) = S_2 \rightarrow RQ(q_1 q_2) = S_1 \cup S_2$$

and

$$\begin{aligned}
T = \{&(mclk, CLK_1' \cup CLK_2', (q_1, q_2), m_1 \wedge m_2, c_1 \wedge c_2, r_1 \wedge r_2, O_1' \cup O_2', Proc_1' \cup Proc_2', \\
&(q_1', q_2')) \mid \\
&(mclk, CLK_1', q_1, m_1, c_1, r_1, O_1', Proc_1', q_1') \in T_1, \\
&(mclk, CLK_2', q_2, m_2, c_2, r_2, O_2', Proc_2', q_2') \in T_2, m_1 \wedge m_2 \neq false, c_1 \wedge c_2 \neq false\}^1
\end{aligned}$$

$$\cup\{(clk_1,\varnothing,(q_1,q_2),m_1,c_1,r_1,O_1',Proc_1',(q_1',q_2))\mid$$
$$(clk_1,\varnothing,q_1,m_1,c_1,r_1,O_1',Proc_1',q_1')\in T_1, clk_1\neq mclk\}^2$$
$$\cup\{(clk_2,\varnothing,(q_1,q_2),m_2,c_2,r_2,O_2',Proc_2',(q_1,q_2'))\mid$$
$$(clk_2,\varnothing,q_2,m_2,c_2,r_2,O_2',Proc_2',q_2')\in T_2, clk_2\neq mclk\}^3$$

1. Both FSMs take their *mclk* driven transitions simultaneously.
2. The first FSM makes a transition driven by a clock different than *mclk*, while the second one is idle.
3. The second FSM makes a transition driven by a clock different than *mclk*, while the first one is idle.

The conditions $m_1 \wedge m_2 \neq false$ and $c_1 \wedge c_2 \neq false$ in the above definition are important since they remove unwanted transitions that arise when the two FSMs have common inputs. For a common input *a*, four combinations are produced: $a.a$, $\bar{a}.a$, $a.\bar{a}$, and $\bar{a}.\bar{a}$. $\bar{a}.a$ and $a.\bar{a}$ must be removed since they have no meaning. They always evaluate to false. They would also be a source of non-determinism since they represent the same monomial. After the removal, the number of transitions is halved. For *n* common inputs, the number of transitions is reduced by 2^n. While only synchronous product appears in the parallel operator of Argos, the synchronous parallel operator of DFCharts involves both synchronous product (transition subset 1) and interleaving (transitions subsets 2 and 3). Subsets 2 and 3 are empty unless the asynchronous parallel operator has been applied beforehand at a lower hierarchical level.

Figures 4.11–4.13 show FSMs A1, A2 and A3, respectively, and are used to illustrate the application of the synchronous parallel operator. A3 is produced

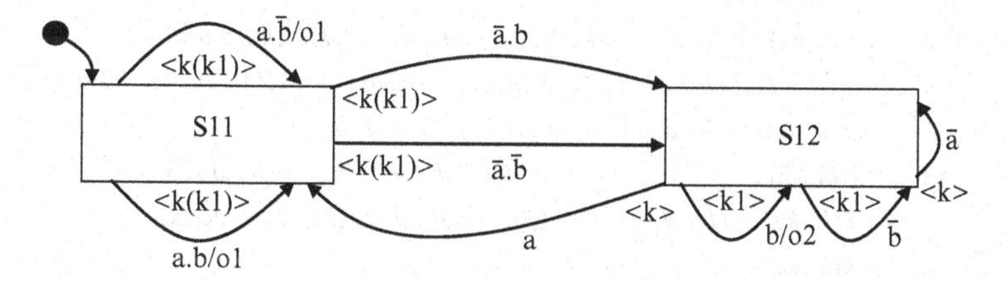

Fig. 4.11 FSM A1

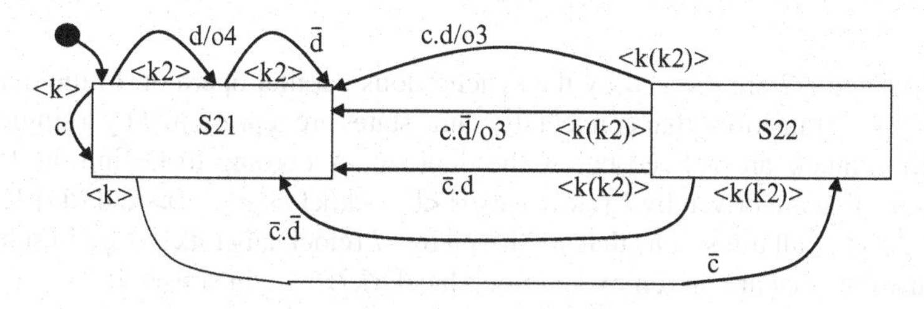

Fig. 4.12 FSM A2

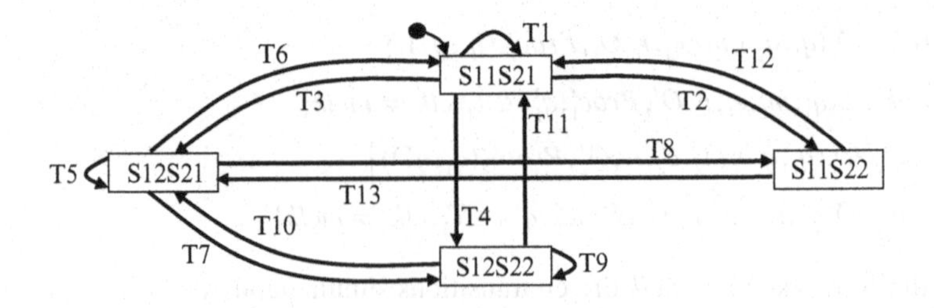

Transitions from S11S21:

$T1 = \{<k2>\overline{d}; <k2> d/o4; <k(k1)> a.\overline{b}.c/o1; <k(k1)> a.b.c/o1\}$

$T2 = \{<k(k1)> a.\overline{b}.\overline{c}/o1; <k(k1)> a.b.\overline{c}/o1\}$

$T3 = \{<k(k1)> \overline{a}.\overline{b}.c; <k(k1)> \overline{a}.b.c\}$

$T4 = \{<k(k1)> \overline{a}.\overline{b}.\overline{c}; <k(k1)> \overline{a}.b.\overline{c}\}$

Transitions from S12S21:

$T5 = \{<k1>\overline{b}; <k1> b/o2; <k2>\overline{d}; <k2> d/o4; <k> \overline{a}.c\}$

$T6 = \{<k> a.c\}$

$T7 = \{<k> \overline{a}.\overline{c}\}$

$T8 = \{<k> a.\overline{c}\}$

Transitions from S12S22:

$T9 = \{<k1>\overline{b}; <k1> b/o2\}$

$T10 = \{<k(k2)> \overline{a}.\overline{c}.\overline{d}; <k(k2)> \overline{a}.\overline{c}.d; <k(k2)> \overline{a}.c.\overline{d}/o3; <k(k2)> \overline{a}.c.d/o3\}$

$T11 = \{<k(k2)> a.\overline{c}.\overline{d}; <k(k2)> a.\overline{c}.d; <k(k2)> a.c.\overline{d}/o3; <k(k2)> a.c.d/o3\}$

Transitions from S11S22:

$T12 = \{<k(k1,k2)> a.\overline{b}.\overline{c}.\overline{d}/o1; <k(k1,k2)> a.\overline{b}.\overline{c}.d/o1; <k(k1,k2)> a.\overline{b}.c.\overline{d}/o1,o3;$

$<k(k1,k2)> a.\overline{b}.c.d/o1,o3; <k(k1,k2)> a.b.\overline{c}.\overline{d}/o1; <k(k1,k2)> a.b.\overline{c}.d/o1;$

$<k(k1,k2)> a.b.c.\overline{d}/o1,o3; <k(k1,k2)> a.b.c.d/o1,o3\}$

$T13 = \{<k(k1,k2)> \overline{a}.\overline{b}.\overline{c}.\overline{d}; <k(k1,k2)> \overline{a}.\overline{b}.\overline{c}.d; <k(k1,k2)> \overline{a}.\overline{b}.c.\overline{d}/o3,$

$<k(k1,k2)> \overline{a}.\overline{b}.c.d/o3; <k(k1,k2)> \overline{a}.b.\overline{c}.\overline{d}; <k(k1,k2)> \overline{a}.b.\overline{c}.d;$

$<k(k1,k2)> \overline{a}.b.c.\overline{d}/o3; <k(k1,k2)> \overline{a}.b.c.d/o3\}$

Fig. 4.13 FSM A3 = A1||A2

when A1 and A2 are merged by the synchronous parallel operator. In the diagram showing A3, transitions that connect the same states are represented by a single line. Transition labels are defined below the diagram. According to Definition 4.7, all transitions that are driven by *k* (clock labels <k>, <k(k1)>, <k(k2)> and <k(k1,k2)>) are in subset 1; all transitions that are driven by *k1* (clock label <k1>) are in subset 2; all transitions that are driven by *k2* (clock label <k2>) are in subset 3.

Theorem 4.1. *When the synchronous parallel operator is applied on two deterministic and reactive FSMs the resulting FSM is deterministic and reactive.*

Proof

mclk: Let I_{q1}, C_{q1}, R_{q1} be the sets of input signals, variable conditions and CS signals bounded to *mclk* respectively for state q_1 and similarly I_{q2}, C_{q2}, R_{q2} for state q_2. Let ci be the number of common inputs (signals and conditions), $ci = |I_{q1} \cap I_{q2}| + |C_{q1} \cap C_{q2}|$. The input signals, variable conditions and CS signals for the equivalent state q_1q_2 are $I_{q1q2} = I_{q1} \cup I_{q2}$, $C_{q1q2} = C_{q1} \cup C_{q2}$ are $R_{q1q2} = R_{q1} \cup R_{q2}$. The total number of inputs is $ni = |I_{q1q2}| + |C_{q1q2}| + |R_{q1q2}|$. A necessary condition for reactivity and determinism in state q_1q_2 is that the number of outgoing transitions is $nt = 2^{ni}$.

From the definition of subset 1 of T, it is evident that the number of outgoing transitions from state q_1q_2 is equal to $nt = (nt_1 \cdot nt_2)/2^{ci}$ where nt_1 and nt_2 are the numbers of outgoing transitions from the states q_1 and q_2 respectively. Since both input FSMs are deterministic and reactive $n_1 = 2^{(|I_{q1}|+|C_{q1}|+|R_{q1}|)}$ and $n_2 = 2^{(|I_{q2}|+|C_{q2}|+|R_{q2}|)}$. Hence, $n = (2^{(|I_{q1}|+|C_{q1}|+|R_{q1}|)} . 2^{(|I_{q2}|+|C_{q2}|+|R_{q2}|)})/2^{ci} = 2^{(|I_{q1}|+|I_{q2}|+|C_{q1}|+|C_{q2}|+|R_{q1}|+|R_{q2}|-ci)} = 2^{ni}$.

For non-determinism to be present, there would have to be two transitions out of q_1q_2 with the same monomials, $m_1.c_1.r_1 \wedge m_2.c_2.r_2 = m_1'.c_1'.r_1' \wedge m_2'.c_2'.r_2'$, which requires $m_1.c_1.r_1 = m_1'.c_1'.r_1'$ and $m_2.c_2.r_2 = m_2'.c_2'.r_2'$. However, this cannot be true since the assumption is that both input FSMs are deterministic. If each transition in T is unique and there are 2^{ni} transitions then all input combinations are defined which means that reactivity is also satisfied.

Other clocks: Transitions that appear in subsets 2 and 3 are simply copied as in Fig. 4.10. Thus determinism and reactivity are preserved. ■

4.1.3 Asynchronous Parallel Operator

In the definition below *CR* is the set of common CS signals between the two FSMs, $CR = R_1 \cap R_2$. CR must not be empty. The left operand has to be a gclk FSM while the right operand has to be a non-gclk FSM. The transition set T is divided into four subsets that are explained below.

Definition 4.8. Asynchronous parallel operator

$$
\begin{aligned}
&(CLK_1, Q_1, q_{01}, I_1, R_1, O_1, V_1, C_1, T_1, RQ_1, Proc_1) \backslash\backslash \\
&(CLK_2, Q_2, q_{02}, I_2, R_2, O_2, V_2, C_2, T_2, RQ_2, Proc_2) \\
&= (CLK_1 \cup CLK_2, Q_1 \times Q_2, (q_{01}, q_{02}), I_1 \cup I_2, (R_1 \cup R_2) \backslash CR, O_1 \cup O_2, \\
&\quad V_1 \cup V_2, C_1 \cup C_2\, T, RQ, Proc_1 \cup Proc_2)
\end{aligned}
$$

where

$$RQ_1(q_1) = S_1 \wedge RQ_2(q_2) = S_2 \rightarrow RQ(q_1 q_2) = (S_1 \cup S_2) \setminus (S_1 \cap S_2)$$

$$\begin{aligned}
T = \{ &(gclk, CLK_1', (q_1, q_2), m_1, c_1, HF(r_1, r_1^a \cap CR), O_1', Proc_1', (q_1', q_2)) \mid \\
&\quad (gclk, CLK_1', q_1, m_1, c_1, r_1, O_1', Proc_1, q_1') \in T_1, \\
&\quad RQ(q_1) \cap RQ(q_2) = \varnothing, r_1^+ \cap CR = \varnothing \}^1 \\
\cup \{ &(gclk, CLK_1' \cup clk_2, (q_1, q_2), m_1 \wedge m_2, c_1 \wedge c_2, HF(r_1 \wedge r_2, (r_1 \wedge r_2)^a \cap CR), \\
&O_1' \cup O_2', Proc_1' \cup Proc_2', (q_1', q_2')) \mid \\
&(gclk, CLK_1', q_1, m_1, c_1, r_1, O_1', Proc_1', q_1') \in T_1, \\
&(clk_2, \varnothing, q_2, m_2, c_2, r_2, O_2', Proc_2', q_2') \in T_2, \\
&RQ(q_1) \cap RQ(q_2) \neq \varnothing, (r_1 \wedge r_2)^+ \cap CR = RQ(q_1) \cap RQ(q_2), r_1 \wedge r_2 \neq false \}^2 \\
\cup \{ &(clk_1, \varnothing, (q_1, q_2), m_1, c_1, r_1, O_1', Proc_1', (q_1', q_2)) \mid \\
&(clk_1, \varnothing, q_1, m_1, c_1, r_1, O_1', Proc_1', q_1') \in T_1 \}^3 \\
\cup \{ &(clk_2, \varnothing, (q_1, q_2), m_2, c_2, HF(r_2, r_2^a \cap CR), O_2', Proc_2', (q_1, q_2')) \mid \\
&(clk_2, \varnothing, q_2, m_2, c_2, r_2, O_2', Proc_2', q_2') \in T_2, RQ(q_1) \cap RQ(q_2) = \varnothing, r_2^+ \cap CR = \varnothing \}^4
\end{aligned}$$

1. The first FSM makes a *gclk* transition while the second one is idle. In the current states q_1 and q_2 the two FSMs have no common CS signals as denoted by $RQ(q_1) \cap RQ(q_2) = \varnothing$. If q_1 has a CS signal that belongs to the CR set, a rendezvous on the corresponding channel will never happen while the equivalent machine is in q_1q_2. Therefore, all transitions where any common CS signal appears as a positive input are removed as denoted by the condition $r_1^+ \cap CR = \varnothing$. Transitions where all common CS signals appear as negative inputs survive, but the common CS signals are removed with the hiding function as denoted by $HF(r_1, r_1^a \cap CR)$.
2. The two FSMs are in the states which share at least one CS signal. ($RQ(q_1) \cap RQ(q_2) \neq \varnothing$). The next ticks of *gclk* and the clock of the second FSM will occur simultaneously. A transition survives if CS signals corresponding to common channels where rendezvous is about to happen (that belong to set $RQ(q_1) \cap RQ(q_2)$) appear as positive inputs. This is denoted by the condition $(r_1 \wedge r_2)^+ \cap CR = RQ(q_1) \cap RQ(q_2)$. This condition implies at the same time that if a CS signal corresponds to a common channel where rendezvous will not happen in the next tick (belongs to set $CR / (RQ(q_1) \cap RQ(q_2))$) it has to appear as a negative input. All common CS signals are removed with the hiding function as denoted by $HF(r_1 \wedge r_2, (r_1 \wedge r_2)^a \cap CR)$. The condition $r_1 \wedge r_2 \neq false$ removes all invalid monomials like $\overline{a}.a$.
3. The first FSM makes a transition not driven by *mclk*. The second one is idle.
4. The second FSM makes a transition while the first one is idle. The two FSMs are in states which don't share any CS signals. Therefore, only transitions where all common CS signals appear as negative inputs survive.

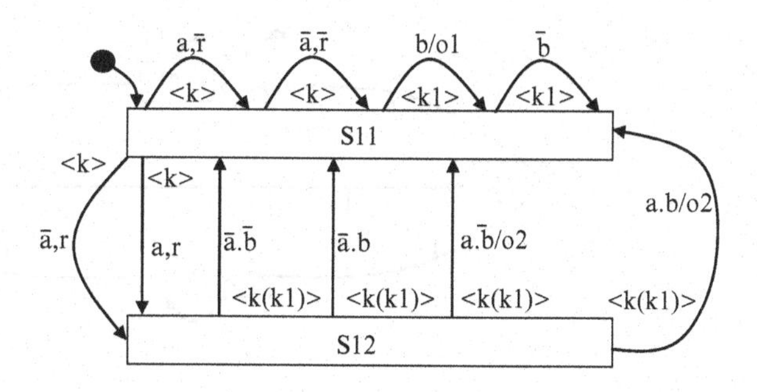

Fig. 4.14 FSM A4

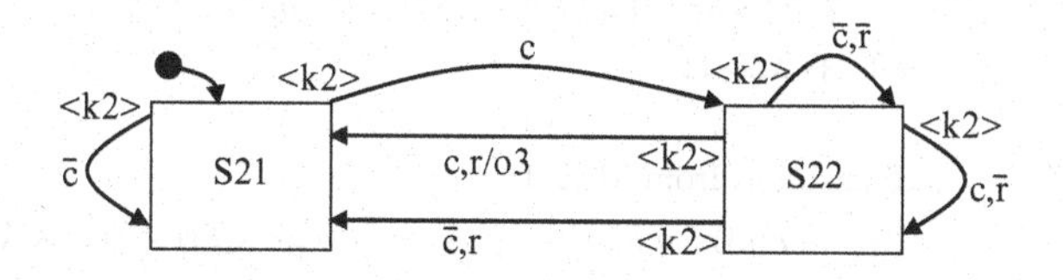

Fig. 4.15 FSM A5

While the synchronous product applied in the synchronous parallel operator involves a single clock, the synchronous product applied in the asynchronous parallel operator involves two different clocks. Both operators use interleaving.

The asynchronous parallel operator in CRSM [92], which combines two single-clock FSMs, distinguishes three cases: the first FSM takes a transition while the second one is idle, the second FSM takes a transition while the first one is idle, and both FSMs take transitions simultaneously. These three cases are also found in the asynchronous composition in CCS [14] and CSP [13]. In DFCharts, one of the input FSMs is multiclock in general. The transitions of the multiclock FSM that are driven by *gclk* have to be treated separately from those that are driven by other clocks, since only gclk-driven transitions are involved in rendezvous. Thus, there are four cases.

Figures 4.14–4.16 show FSMs A4, A5 and A6, respectively, which are used to illustrate the application of the asynchronous parallel operator. In Figs. 4.14 and 4.15, *r* is a CS signal. It is separated from input signals by "," so that it is easy to distinguish. Comma means AND, exactly like dot. A6 is produced when A4 and A5 are merged by the asynchronous parallel operator. According to Definition 4.8 transitions in A6 can be classified as follows: all transitions that are driven by *k*, where *k* is not synchronized with *k2* (clock labels <k> and <k(k1)>), are in subset 1; all transitions that are driven by *k*, where *k* is synchronized with *k2* (clock label <k(k2)>), are in subset 2; all transitions that are driven by *k1* (clock label <k1>) are in subset 3; all transitions that are driven by *k2* (clock label <k2>) are in subset 4.

We look at how common CS signals are removed by the asynchronous parallel operator by considering two states of the equivalent FSM A6 in Fig. 4.16, S11S21 where there is no rendezvous, and S11S22 where rendezvous occurs. There are four transitions driven by *k* out of state S11 with monomials $\overline{a}.\overline{r}$, $a.\overline{r}$, $\overline{a}.r$, $a.r$.

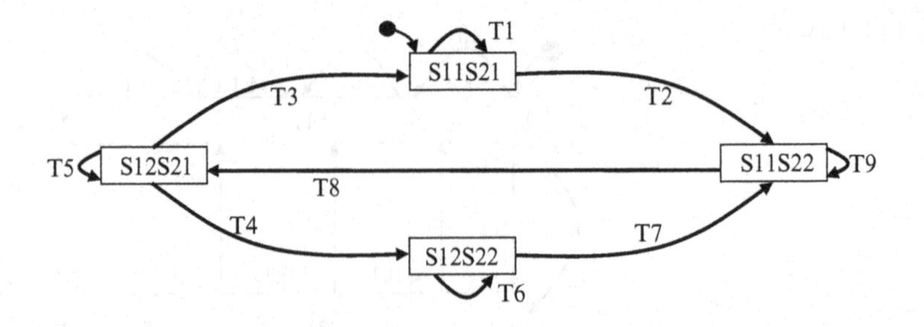

Transitions from S11S21:

$T1 = \{<k1>\bar{b};<k1>b/o1;<k>\bar{a};<k>a;<k2>\bar{c}\}$

$T2 = \{<k2>c\}$

Transitions from S12S21:

$T3 = \{<k(k1)>a.b/o2;<k(k1)>a.\bar{b}/o2;<k(k1)>\bar{a}.b;<k(k1)>\bar{a}.\bar{b}\}$

$T4 = \{<k2>c\}$

$T5 = \{<k2>\bar{c}\}$

Transitions from S12S22:

$T6 = \{<k2>c,<k2>\bar{c}\}$

$T7 = \{<k(k1)>a.b/o2;<k(k1)>a.\bar{b}/o2;<k(k1)>\bar{a}.b;<k(k1)>\bar{a}.\bar{b}\}$

Transitions from S11S22:

$T8 = \{<k(k2)>\bar{a}.\bar{c};<k(k2)>\bar{a}.c/o3;<k(k2)>a.\bar{c};<k(k2)>a.c/o3\}$

$T9 = \{<k1>\bar{b};<k1>b/o1\}$

Fig. 4.16 FSM A6

When S11 is combined with S21, transitions where r appears as a positive input have to be removed, which leaves $\bar{a}.\bar{r}$ and $a.\bar{r}$, and after hiding $\bar{a}$ and a. The transitions that are driven by *k1* and *k2* are simply copied.

CS signal r is common for states S11 and S22. Their outgoing transitions driven by *k* and *k2* are taken simultaneously because of rendezvous. There are 16 such transitions initially, but after conditions $(r_1 \wedge r_2)^+ \cap CR = RQ(q_1) \cap RQ(q_2)$ and $r_1 \wedge r_2 \neq false$ are applied only four survive as shown in Fig. 4.16. The transitions that are driven by *k1* are copied.

For each channel, it is known at compile-time in which state the rendezvous will happen. This is possible due to the absence of strong abort in DFCharts. For example, once FSM A6 enters state S11S22 the rendezvous on channel r must happen in the next tick of *k*. If a weak abort occurs in the next tick due to a higher level transition, A6 will be allowed to react. Due to the way in which FSMs are composed in DFCharts, A6 can only be preempted by a transition that is driven by *k*.

Theorem 4.2. *When the asynchronous parallel operator is applied on two deterministic and reactive FSMs, the resulting FSM is deterministic and reactive.*

Proof

mclk: When the equivalent FSM is in a state where rendezvous will occur in the next tick, mclk transitions are synchronously combined with the transitions of the other clock. It was already proved that synchronous product preserves determinism and reactivity. When a common CS signal is removed, other inputs are not affected. For a CS signal r and monomial m, there are initially two possibilities: $\overline{r}.m$ and $r.m$. If a positive CS signal is removed, $\overline{r}.m$ remains and after hiding only m. If a negative CS signal is removed, $r.m$ remains and after hiding only m. In both cases, reactivity and determinism are preserved.

When the equivalent FSM is in a state where rendezvous cannot occur in the next tick, *mclk* driven transitions are copied. The removal of negative CS inputs does not affect determinism and reactivity as shown above.

Other clocks: The same argument applies as in the second paragraph in the *mclk* part. Transitions are copied, while removal of negative CS signals does not affect determinism and reactivity.■

4.1.4 Hiding Operator

Definition 4.9. Hiding/localization operator

$$(CLK,Q,q_0,I,R,O,V,C,T,RQ,Proc)\backslash a=(CLK,Q,q_0,I\backslash a,R,O\backslash a,V,C,T',RQ,Proc)$$

where

$$T'=\{(clk,CLK',q,HF(m,\{a\}),c,r,O'\backslash a,Proc',q')\mid (clk,CLK',q,m,c,r,O',Proc',q')\in T,a\in O',a\in m^{+}\}^{1}$$
$$\cup\{(clk,CLK',q,HF(m,\{a\}),c,r,O',Proc',q')\; (clk,CLK',q,m,c,r,O',Proc',q')\in T,a\notin O',a\in m^{-}\}^{2}$$

1. If the local signal is present in the input part of the transition and emitted in the output part of the transition, then the transition survives and the local signal is hidden.
2. If the local signal is absent in the input part of the transition and not emitted in the output part of the transition, then the transition survives and the local signal is hidden.

Hiding was already seen in the context of the asynchronous parallel operator. That type of hiding should not be confused with the one defined above. The hiding operator from Definition 4.9 is used to synchronize two FSMs that are connected by the synchronous parallel operator or refinement operator. The synchronization is performed with local signals. In contrast, the other type of hiding is built into the asynchronous parallel operator and works with CS signals. The presence of two types of hiding is not surprising as DFCharts employs two communication mechanisms, synchronous broadcast and rendezvous.

The hiding operator from the definition above preserves neither determinism nor reactivity. The causality cycles that can appear are exactly the same as those described in the Argos section.

4.1.5 Refinement Operator

There are two variations of the refinement operator. Consequently, the definition below consists of two parts. The first type of the refinement operator, labelled $\downarrow$, is used when both operands are either gclk FSMs or non-gclk FSMs. The second, labelled $\Downarrow$, is used when the left operand is a gclk FSM while the right operand is a non-gclk FSM. FSMs connected by the synchronous parallel operator must not have any common CS signals.

Definition 4.10. Refinement operator

$$
\begin{aligned}
&(CLK_1, Q_1, q_{01}, I_1, R_1, O_1, V_1, C_1, T_1, RQ_1, Proc_1) \downarrow_q \\
&\quad (CLK_2, Q_2, q_{02}, I_2, R_2, O_2, V_2, C_2, T_2, RQ_2, Proc_2) \\
&\quad = (CLK_1 \cup CLK_2, (Q_1 \setminus \{q\}) \cup (q \times Q_2), q_{01}, I_1 \cup I_2, R_1 \cup R_2, O_1 \cup O_2, \\
&\qquad V_1 \cup V_2, C_1 \cup C_2 T, RQ, Proc_1 \cup Proc_2)
\end{aligned}
$$

where

$$RQ_1(q_1) = S_1 \wedge RQ_2(q_2) = S_2 \rightarrow RQ(q_1 q_2) = S_1 \cup S_2$$

and

$$
\begin{aligned}
T = &\{(mclk, CLK_1', q_1, m_1, c_1, r_1, O_1', Proc_1', q_2) \in T_1 \mid q_1, q_2 \neq q\}^1 \\
&\cup \{(mclk, CLK_1', q_1, m_1, c_1, r_1, O_1', Proc_1', (q, q_{02})) \mid \\
&\quad (mclk, CLK_1', q_1, m_1, c_1, r_1, O_1', Proc_1', q) \in T_1\}^2 \\
&\cup \{(mclk, CLK_2', (q, q_1), m_1 \wedge m_2, c_1 \wedge c_2, r_2, O_2', Proc_2', (q, q_2)) \mid \\
&\quad (mclk, \varnothing, q, m_1, c_1, -, \varnothing, \varnothing, q) \in T_1, m_1^+ = \varnothing, \\
&\quad (mclk, CLK_2', q_1, m_2, c_2, r_2, O_2', Proc_2', q_2) \in T_2, m_1 \wedge m_2 \neq false, c_1 \wedge c_2 \neq false\}^3 \\
&\cup \{(mclk, CLK_2', (q, q_1), m_1 \wedge m_2, c_1 \wedge c_2, r_2, O_1' \cup O_2', Proc_1' \cup Proc_2', (q, q_{02})) \mid \\
&\quad (mclk, \varnothing, q, m_1, c_1, -, O_1', Proc_1', q) \in T_1, m_1^+ \neq \varnothing,
\end{aligned}
$$

$$(mclk, CLK_2', q_1, m_2, c_2, r_2, O_2', Proc_2', q_2) \in T_2, m_1 \wedge m_2 \neq false, c_1 \wedge c_2 \neq false\}^4$$

$$\cup\{(mclk, CLK_2', (q, q_1), m_1 \wedge m_2, c_1 \wedge c_2, r_2, O_1' \cup O_2', Proc_1' \cup Proc_2', q') \mid$$

$$(mclk, \varnothing, q, m_1, c_1, -, O_1', Proc_1', q') \in T_1,$$

$$(mclk, CLK_2', q_1, m_2, c_2, r_2, O_2', Proc_2', q_2) \in T_2, m_1 \wedge m_2 \neq false, c_1 \wedge c_2 \neq false\}^5$$

$$\cup\{(clk_1, \varnothing, q_1, m_1, c_1, r_1, O_1', Proc_1', q_2) \in T_1 \mid q_1, q_2 \neq q, clk_1 \neq mclk\}^6$$

$$\cup\{(clk_2, \varnothing, (q, q_1), m_2, c_2, r_2, O_2', Proc_2', (q, q_2)) \mid$$

$$(clk_2, \varnothing, q_1, m_2, c_2, r_2, O_2', Proc_2', q_2) \in T_2\}^7$$

As in the definition of the synchronous parallel operator, *mclk* stands for *gclk* when $\downarrow$ is applied on two gclk FSMs. In the case of non-gclk FSMs, *mclk* represents the single clock of the two FSMs. The transition set, which consists of seven subsets, is explained below.

1. *mclk* driven transitions of the first FSM that don't involve the refined state.
2. *mclk* driven transitions of the first FSM that enter the refined state.
3. The second (refining) FSM makes a mclk-driven transition while the first FSM makes a mclk transition that loops back to the refined state. However, the second FSM is not reset to its initial state q_{02} since the transition of the first FSM does not involve any inputs that must be present ($m_1^+ = \varnothing$).Transitions where all inputs are absent are not specified by the programmer, they are inserted in order to obtain complete monomials. So they should not reset the refining FSM.
4. Same as 3, but this time at least one of the inputs in the first FSM's transition is present, so the refining FSM is reset to its initial state.
5. The refined state is left. Since the abort is weak, the outputs of the second FSM are produced.
6. The first FSM makes the transition that is not driven by *mclk*. The refined state is not involved, since it is always entered and exited by mclk-driven transitions.
7. The second FSM makes the transition that is not driven by *mclk* so the first FSM is idle.

The second part of the definition is used when the state of a gclk FSM is refined by a non-gclk FSM.

$$(CLK_1, Q_1, q_{01}, I_1, R_1, O_1, V_1, C_1, T_1, RQ_1, Proc_1) \Downarrow_q$$
$$(CLK_2, Q_2, q_{02}, I_2, R_2, O_2, V_2, C_2, T_2, RQ_2, Proc_2)$$
$$= (CLK_1 \cup CLK_2, (Q_1 \setminus \{q\}) \cup (q \times Q_2), q_{01}, I_1 \cup I_2, R_1 \cup R_2, O_1 \cup O_2,$$
$$V_1 \cup V_2, C_1 \cup C_2\ T, RQ, Proc_1 \cup Proc_2)$$

where

$$RQ_1(q_1) = S_1 \wedge RQ_2(q_2) = S_2 \rightarrow RQ(q_1 q_2) = S_1 \cup S_2$$

and

$$
\begin{aligned}
T = & \{(gclk, CLK_1', q_1, m_1, c_1, r_1, O_1', Proc_1', q_2) \in T_1 \mid q_1, q_2 \neq q\}^1 \\
& \cup \{(gclk, CLK_1', q_1, m_1, c_1, r_1, O_1', Proc_1', (q, q_{02})) \mid \\
& \quad (gclk, CLK_1', q_1, m_1, c_1, r_1, O_1', Proc_1', q) \in T_1\}^2 \\
& \cup \{(gclk, \varnothing, (q, q_1), m_1, c_1, -, \varnothing, \varnothing, (q, q_1)) \mid \\
& \quad (gclk, \varnothing, q, m_1, c_1, -, \varnothing_1, \varnothing, q) \in T_1, m_1^+ = \varnothing\}^3 \\
& \cup \{(gclk, \varnothing, (q, q_1), m_1, c_1, -, O_1', Proc_1', (q, q_{02})) \mid \\
& \quad (gclk, \varnothing, q, m_1, c_1, -, O_1', Proc_1', q) \in T_1, m_1^+ \neq \varnothing\}^4 \\
& \cup \{(gclk, \varnothing, (q, q_1), m_1, c_1, -, O_1', Proc_1', q_2) \mid \\
& \quad (gclk, \varnothing, q, m_1, c_1, -, O_1', Proc_1', q_2) \in T_1\}^5 \\
& \cup \{(clk_1, \varnothing, q_1, m_1, c_1, r_1, O_1', Proc_1', q_2) \in T_1 \mid q_1, q_2 \neq q, clk_1 \neq mclk\}^6 \\
& \cup \{(clk_2, \varnothing, (q, q_1), m_2, c_2, r_2, O_2', Proc_2', (q, q_2)) \mid \\
& \quad (clk_2, \varnothing, q_1, m_2, c_2, r_2, O_2', Proc_2', q_2) \in T_2\}^7
\end{aligned}
$$

The transition set, which consists of seven subsets, is explained below.

1. *gclk* driven transitions of the first FSM that don't involve the refined state.
2. Transitions of the first FSM that enter the refined state.
3. The first FSM makes a transition that loops back to the refined state. The second FSM is not reset to its initial state q_{02} since the transition of the first FSM does not involve any input signals that must be present ($m_1^+ = \varnothing$).
4. Same as 3, but this time at least one of the input signals in the first FSM's transition is present, so the refining FSM is reset to its initial state.
5. The refined state is left. The abort is weak but there are no outputs from the second FSM, since it does not react in this tick.
6. The first FSM makes a transition that is not driven by *gclk*. The refined state is not involved, since it is always entered and exited by mclk-driven transitions.
7. The second FSM makes a transition while the first FSM is idle.

The definitions of the two versions of the hierarchical operator ($\downarrow$ and $\Downarrow$) are similar but there is an important difference, which creates the need for having two versions instead of just one. In the definition of $\downarrow$, synchronous product is present in subsets 3, 4 and 5 of T. In the definitions of $\Downarrow$, synchronous product is completely absent. Transitions of the two input FSMs are just interleaved.

Figures 4.17–4.19 show FSMs A7, A8 and A9, respectively, which are used to illustrate the application of the second version of the hierarchical operator. A9 is produced when A8 refines state S12 of A7.

The first version of the hierarchical operator is illustrated with A9, A10 shown in Fig. 4.20, and A11 shown in Fig. 4.21. A11 is produced when state S11 of A9 is refined by A10.

Fig. 4.17 FSM A7

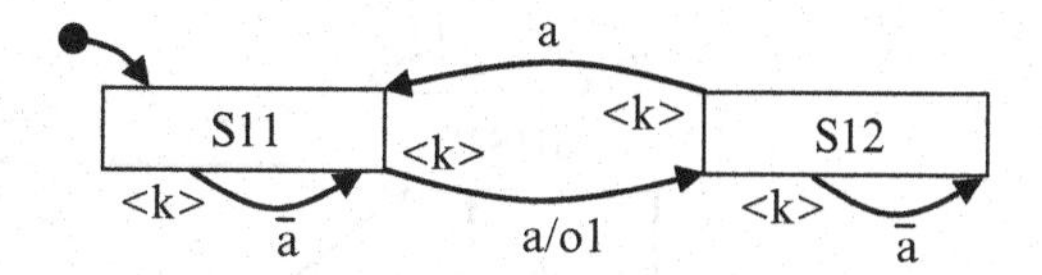

Fig. 4.18 FSM A8

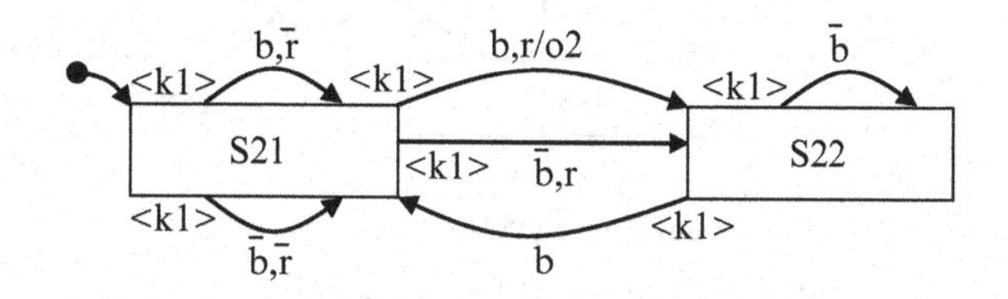

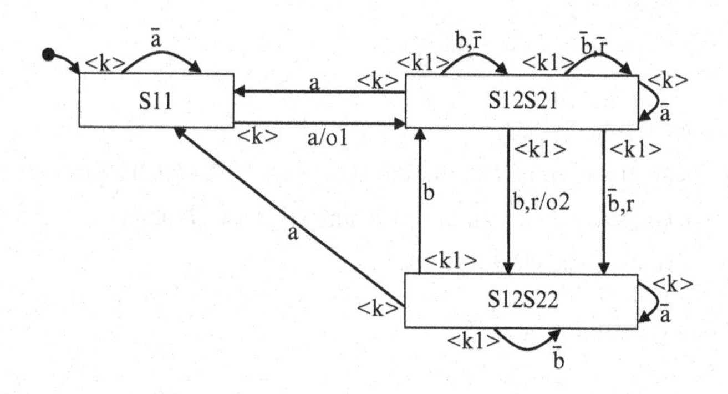

Fig. 4.19 FSM A9=A7 $\Downarrow_{S12}$ A8

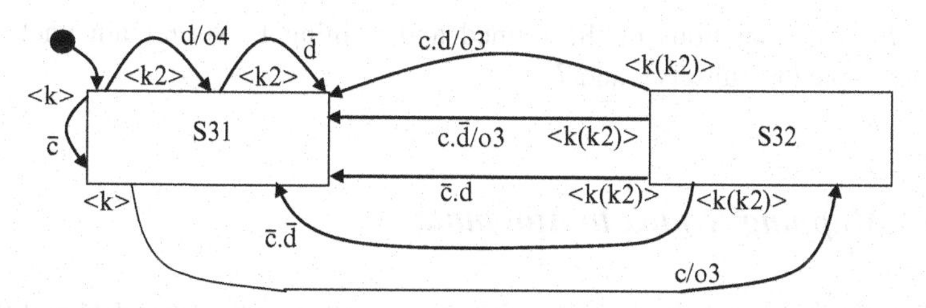

Fig. 4.20 FSM A10

Theorem 4.3. *When the refinement operator is applied on two deterministic and reactive FSMs, the resulting FSM is deterministic and reactive.*

Proof

mclk: In the second version of the hierarchical operator ($\Downarrow$) all *mclk* transitions are copied from the refined FSM. In the first version ($\downarrow$), this is the case in subsets 1 and 2 of *T*. In subsets 3, 4 and 5, mclk transitions of the refined FSM are synchronously

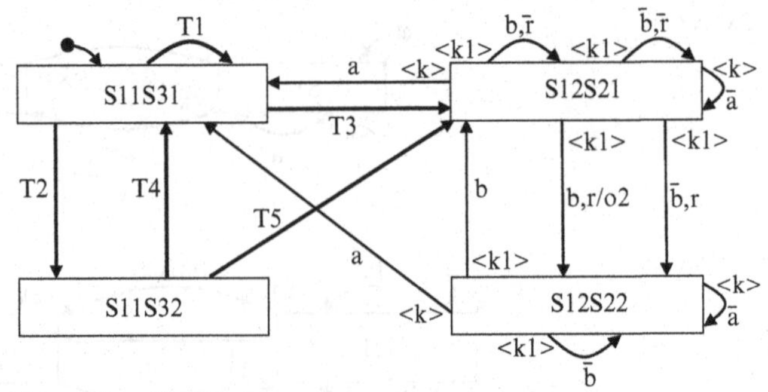

Transitions from S11S31:

$T1 = \{<k2> d/o4; <k2> \bar{d}; <k> \bar{a}.\bar{c}\}$

$T2 = \{<k> \bar{a}.c/o3\}$

$T3 = \{<k> a.\bar{c}/o1; <k> a.c/o1,o3\}$

Transitions from S11S32:

$T4 = \{<k(k2)> \bar{a}.\bar{c}.\bar{d}; <k(k2)> \bar{a}.\bar{c}.d; <k(k2)> \bar{a}.c.\bar{d}/o3; <k(k2)> \bar{a}.c.d/o3\}$

$T5 = \{<k(k2)> a.\bar{c}.\bar{d}/o1; <k(k2)> a.\bar{c}.d/o1; <k(k2)> a.c.\bar{d}/o1,o3;$
$<k(k2)> a.c.d/o1,o3\}$

Fig. 4.21 FSM A11 = A9 $\downarrow_{S11}$ A10

combined with mclk transitions of the refining FSM. It was already proved for Theorem 3.1 that synchronous product preserves reactivity and determinism.

Other clocks: Transitions of the refined and refining FSMs remain unchanged. They are copied to subsets 6 and 7.

4.1.6 Mapping Syntax to Automata

We identify two syntactic domains, the set of gclk specifications *PM* and the set of non-gclk specifications *PS*. Two special variables *M* and *S* will be used to stand for elements of *PM* and *PS*, respectively. There are also two semantic domains discussed in the previous section, the set of gclk FSMs *AM* and the set of non-gclk FSMs *AS*. Finally, two semantic functions, $MSE : PM \rightarrow AM^{rd} \cup \{\perp\}$ and $SSE : PS \rightarrow AS^{rd} \cup \{\perp\}$, are needed to map syntactic objects to semantic objects. AM^{rd} is the set of deterministic and reactive FSMs within *AM* while AS^{rd} is the set of deterministic and reactive FSMs within *AS*. Specifications that do not meet the requirements for determinism and reactivity are considered incorrect and mapped to $\perp$, pronounced "bottom". Non-determinism and non-reactivity can be produced with the hiding operator.

Listed below are the production rules which show ways in which syntactic objects may be combined. On the syntactic side the synchronous parallel, asynchronous parallel, refinement and localization/hiding operations are labelled $\bar{s}$, $\bar{a}$, $\bar{r}$ and $\bar{l}$.

$$M ::= M\bar{s}M|M\bar{a}S|M\bar{r}(q_1 R_1, q_2 R_2, ..., q_n R_n)|M\bar{l}(a_1, a_2, ..., a_n)$$

$$S ::= S\bar{s}S|S\bar{r}(q_1 S_1, q_2 S_2, ..., q_n S_n)|S\bar{l}(a_1, a_2, ..., a_n)$$

$$R ::= M \mid S$$

All refinements can be specified in a single statement. On the other hand, the refinement operator is defined to work with the refinement of a single state only. Therefore, when giving the meaning to the refinement statement, the refinement operator needs to be applied repeatedly, as many times as there are refinements, in order to reach the final equivalent FSM. Note that subscript n in the refinement statement denotes the number of refined states, not the total number of states, since some states may be unrefined. The localization statement can also specify all local signals at once. Thus, as in the case of the refinement operator, the localization operator may need to be used multiple times to produce the FSM the localization statement is mapped to.

The syntax is mapped to semantics as follows:

$$MSE(M_1\bar{s}M_2) = \begin{cases} \bot \ if \ MSE(M_1) = \bot \ or \ MSE(M_2) = \bot \\ MSE(M_1) \,||\, MSE(M_2) \ otherwise \end{cases}$$

$$MSE(M\bar{a}S) = \begin{cases} \bot \ if \ MSE(M) = \bot \ or \ SSE(S) = \bot \\ MSE(M) \setminus\setminus SSE(S) \ otherwise \end{cases}$$

$$MSE(M\bar{r}(q_1 R_1, q_2 R_2, ..., q_n R_n)) = \begin{cases} \bot \ if \ MSE(M) = \bot \ or \ \exists i \in [1,n] : SE(R_i) = \bot \\ X_n : X_1 = \begin{cases} MSE(M) \downarrow_{q_1} MSE(R_1) \ if \ R ::= M \\ MSE(M) \Downarrow_{q_1} SSE(R_1) \ if \ R ::= S \end{cases} \\ \quad X_i = \begin{cases} X_{i-1} \downarrow_{q_i} MSE(R_i) \ if \ R ::= M \\ X_{i-1} \Downarrow_{q_i} SSE(R_i) \ if \ R ::= S \end{cases} \\ \quad for \ i \in [2,n] \end{cases}$$

$$MSE(M\bar{l}(a_1, a_2, ..., a_n)) = \begin{cases} \bot \ if \ MSE(M) = \bot \\ X_n \ if \ X_n \in AM^{rd} \ else \bot \ otherwise \\ \quad where X_n is defined by: \\ \quad X_1 = MSE(M) \setminus a_1, X_i = X_{i-1} \setminus a_i \ for \ i \in [2,n] \end{cases}$$

$$SSE(S_1 \overline{s} S_2) = \begin{cases} \perp \ if \ \ SSE(S_1) = \perp \ or \ SSE(S_2) = \perp \\ SSE(S_1) \,||\, SSE(S_2) \ \ otherwise \end{cases}$$

$$SSE(S\overline{r}(q_1 S_1, q_2 S_2, \ldots, q_n S_n)) = \begin{cases} \perp \ if \ \ SSE(S) = \perp \ or \ \exists i \in [1,n] : SSE(S_i) = \perp \\ X_n : X_1 = SSE(S) \downarrow_{q_1} \ SSE(S_1), \\ \quad X_i = X_{i-1} \downarrow_{q_i} \ SSE(S_i) \, for \, i \in [2,n] \\ \quad otherwise \end{cases}$$

$$SSE(S\overline{l}\,(a_1, a_2, \ldots, a_n)) \begin{cases} \perp \ if \ \ SSE(S) = \perp \\ X_n \ if \ X_n \in AS^{rd} \ else \perp \ otherwise \\ \quad where \, X_n \, is \, defined \, by: \\ \quad X_1 = SSE(S) \setminus a_1, X_i = X_{i-1} \setminus a_i \ \ for \, i \in [2,n] \end{cases}$$

4.1.7 Integrating SDF Graphs into Automata Semantics

Denotational semantics is the most common means of formally describing Kahn process networks. As a restriction of Kahn process networks, the family of dataflow process network models, which includes SDF, is also captured by denotational semantics. It would be very difficult to integrate denotationally described SDFGs with operationally described synchronous FSMs. Therefore, we have to represent the SDFG as an FSM using the operators defined in previous sections. In every iteration, the SDFG goes through two phases. In the first phase inputs are received and outputs from the previous iteration are sent. In the second phase inputs are processed and outputs are produced. The parallel and refinement operators.

Definition 4.11. Synchronous dataflow graph with m inputs and n outputs

$$SDF = TOP \downarrow_{io} \ IOPAR$$

$$TOP = (Q, q_0, I, R, O, V, C, T, RQ, Proc)$$

where

$Q = (io, processing)$, $q_0 = io$,
$I = \{gs\} \cup \{in\_qi_i \mid 1 \le i \le m\} \cup \{in\_qo_i \mid 1 \le i \le n\}$, $R = \emptyset$, $O = \emptyset$,
$V = \{data\_in_i \mid 1 \le i \le m\} \cup \{data\_out_i \mid 1 \le i \le n\}$, $C = \emptyset$, $T = \{t_1, t_2\}$,
$RQ = \emptyset$, $Proc = \{assign\_output\_values\}$

$$t_1 = (sdfclk, \emptyset, io, in\_qi_1.in\_qi_2 \ldots in\_qi_m.in\_qo_1.in\_qo_2 \ldots \\ in\_qo_n, -, -, \emptyset, \emptyset, \ \ processing)$$

$$t_2 = (sdfclk, \varnothing, processing, gs, -, -, \varnothing, \{assign\_output\_values\}, io)$$

$$IOPAR = IN_1 \parallel IN_2 \parallel ... \parallel IN_m \parallel OUT_1 \parallel OUT_2 \parallel ... \parallel OUT_n$$

$$IN = (Q, q_0, I, R, O, V, C, T, RQ, Proc)$$

where

$Q = (q_1, q_2)$, $q_0 = q_1, I = \{\varnothing\}$, $R = \{ch\_in\}$, $O = \varnothing$, $V = \{data\_in\}$, $T = \{t_1\}$,

$RQ = \{(q_1, ch\_in)\}$, $Proc = \{ch\_out\_proc\}$

$t_1 = (sdfclk, \varnothing, q_1, -, -, ch\_in, \varnothing, ch\_in\_proc, q_2)$

$OUT = (Q, q_0, I, R, O, V, C, T, RQ, Proc)$

where

$Q = (q_1, q_2)$ $q_0 = q_1, I = \{\varnothing\}$, $R = \{ch\_out\}$, $O = \varnothing, V = \{data\_out\}$, $T = \{t_1\}$,

$RQ = \{(q_1, ch\_out)\}$, $Proc = \{ch\_out\_proc\}$

$t_1 = (sdfclk, \varnothing, q_1, -, -, ch\_out, \varnothing, ch\_out\_proc, q_2)$

With this definition, an SDFG can be placed in the AS set (non-gclk FSMs) and used as the right operand of the asynchronous parallel operator. Figure 4.22 shows how an SDF graph with two inputs and two outputs is seen as an FSM. All transitions are driven by *sdfclk*. *din* and *dout* stand for *data_in* and *data_out* in Definition 4.11. *aov* stands for *assign_output_values*. The two phases are represented as two states in the top level FSM. The *io* state is refined by concurrent FSMs that handle receiving and sending data. Note that the output FSMs send the initial values of their *data_out* variables before the first iteration. In the *processing* state *data_in* variables

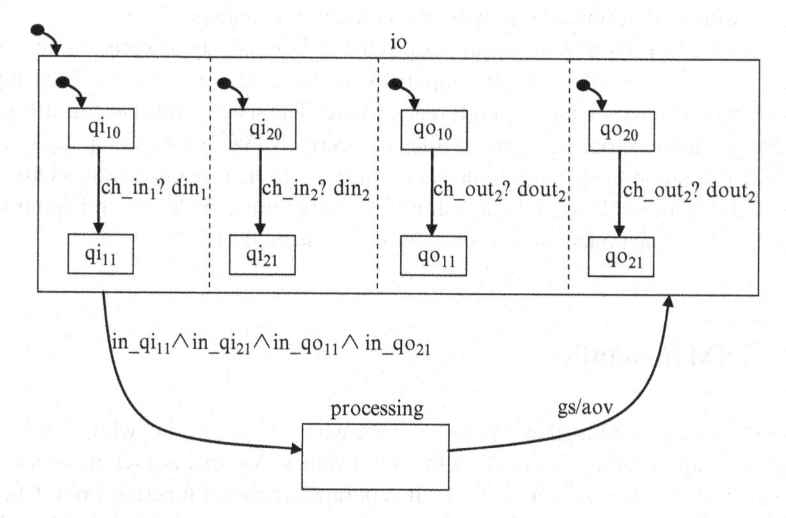

Fig. 4.22 FSM that represents SDF graph with two inputs and two outputs

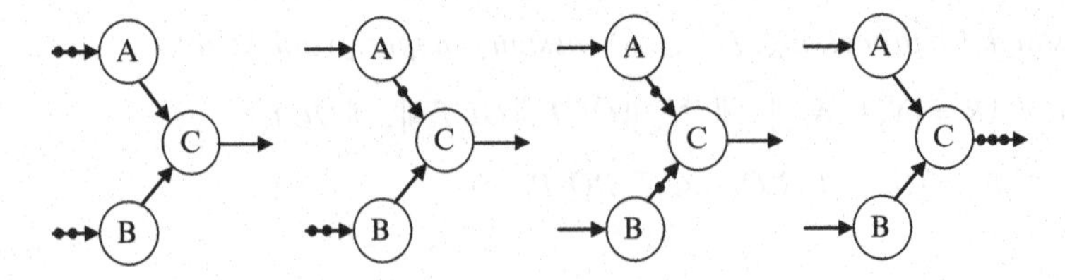

Fig. 4.23 Execution states of SDF1 from Fig. 3.5

are read and *data_out* variables are written. Inputs of type *in_qi* or *in_qo* are present when their corresponding states are active. *Gs* is the signal that denotes that the outputs have been produced leading to the I/O state.

An important feature of "SDF FSM" is that waiting for rendezvous can never be pre-empted on any channel. This significantly simplifies the implementation of rendezvous, which is described in Chap. 7 among other implementation aspects of DFCharts. The implementation of rendezvous is more difficult in a more general case where both sides are allowed to abort [93].

Obviously this is a very high level view as all internal signals are hidden inside the processing state. As we mentioned above, what happens inside the processing state is usually described with denotational semantics. However, Kahn process networks and other dataflow models can also be described operationally as in [39], for example. When the operational semantics is employed, the whole dataflow network is typically seen as a labelled transition system where the state is determined by the state of processes and the state of channels. The state of a channel is simply the number of tokens present in it. In Kahn process networks and other dynamic dataflow models, channels can grow infinitely large resulting in the infinite state space. On the other hand, channels are bounded in SDF. Furthermore, SDF processes contain no state; they transform data in the same way in each firing. Therefore the state of an SDF network is determined purely by the content of channels.

Figure 4.23 shows how an iteration of SDF1 in Fig. 3.5 can be seen as a series of states. The first state is when two input tokens have been received in each input channel which makes actors A and B ready to fire. The second, third and fourth state are entered after actors A, B and C fire, respectively. We need to emphasize that while an execution order has a finite state space, multiple execution orders exist for an SDFG in general. Thus the processing state is not refined in the general definition, but it can be refined once the execution order has been decided.

4.2 TSM Semantics

In TSM an event is defined as a pair $e = (t, v)$, where $t \in T$ is a tag while $v \in V$ is a value. T is the set of tags while V is the set of values. A signal is a set of events. It is a member of $2^{T \times V}$ or a subset of $T \times V$. It is defined as partial function from T to V. Partial in this case means that a signal does not have to be defined for every $t \in T$.

Tags do not necessarily represent physical time. They might only be used to denote the ordering of events. This is the case in DFCharts which is an untimed model. The order among events is described by the ordering relation "≤", which is defined on the set T [25]. The relation is reflexive ($\forall a \in T : a \leq a$), antisymmetric ($\forall a,b \in T : (a \leq b \wedge b \leq a \rightarrow a = b)$), transitive ($\forall a,b,c \in T : (a \leq b \wedge b \leq c \rightarrow a \leq c)$). A related irreflexive relation is also defined, denoted "<". The ordering of tags induces the ordering of events. Given two events $e_1 = (t_1, v_1)$ and $e_1 = (t_2, v_2)$, then $e_1 < e_2$ if and only if $t_1 < t_2$. Two events can also have equivalent tags. In that case they are synchronous.

Since DFCharts is a heterogeneous model that consists of HCFSM and SDF models, we use two sets of tags T' and T'' in order to distinguish between the two constituent models. T' will be used to define synchronous signals in the HCFSM model, while T'' will be used to define asynchronous signals in the SDF model. T' is a totally ordered set, which means that for any two tags t_1' and t_2', $t_1' = t_2'$, $t_1' < t_2'$ or $t_2' < t_1'$. T'' is a partially ordered set. This means that there exist two tags $t_1'', t_2'' \in T''$ such that $t_1'' \nleq t_2''$.

We also define two sets of values V' and V''. V' is the set of values produced in the HCFSM (synchronous) domain while V'' is the set of values produced in the SDF (asynchronous) domain. An event from the synchronous domain is defined as $e' = (t', vec)$ where $t' \in T'$. As indicated in the previous section a valued event in DFCharts carries a group (vector) of values. For this reason we define vec as a tuple of n values which can be taken either from V' or V''. The interface processes, which will be defined shortly, transport values across the two domains. In addition vec can take two special values $\perp$ and p. $\perp$ denotes the absence of an event, while p denotes that the event is present but it carries no value i.e. it is a *pure* event. The set of n-tuples over V' is denoted as V'^n while the set of n-tuples over V'' is denoted as V''^n. Thus $vec \in V'^n \cup V''^n \cup \{\perp, p)$.

Events in the asynchronous SDF domain carry only one value. An event in the asynchronous domain is defined as $e'' = (t'', v)$, where $t'' \in T''$ and $v \in V'' \cup V'$.

The communication between an SDF graph and FSMs is performed using two interface processes shown in Figs. 4.24. and 4.25. *STF* converts SDF output signals into FSM input signals. *FTS* does the opposite. It converts FSM output signals into

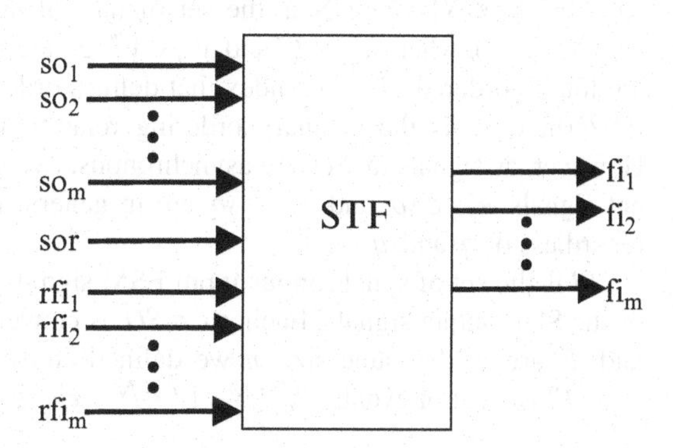

Fig. 4.24 STF process

Fig. 4.25 FTS process

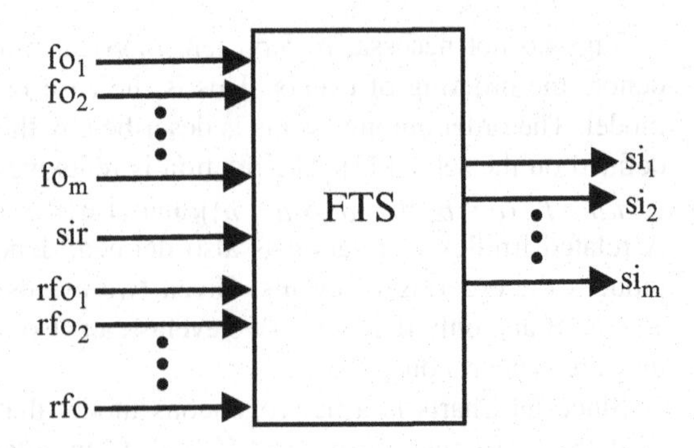

SDF input signals. Note that these processes are implicit, as they are only used for semantic purposes. They are not seen in specifications such as that in Fig. 3.5 where input and output signals in both directions are simply connected together. Both processes are triggered by the same tick signal that is used to trigger all FSMs in a DFCharts specification. When they consume inputs, they instantaneously produce outputs. Thus we can say that they belong to the synchronous domain of DFCharts. When triggered by a tick, each process invokes a set of functions that map parts of input signals to parts of output signals. Each function handles one output signal. Complete output signals are built as processes that are repeatedly triggered by the tick signal.

4.2.1 Data Transfer from SDF to FSM

STF has three inputs: the set of signals *SO*, the set of signals *RFI* and the signal *sor*. It outputs the set of signals *FI*. The signals in *SO* and *FI* are explicit in specification while *sor* and the signals in *RFI* are implicit and should be generated during implementation. We first describe the explicit signals.

$SO = \{so_i \mid 1 \le i \le m\}$ is the set of m output signals of the SDFG that communicate with FSMs. When defined in the TSM framework each $so_i \in SO$ is a set of events $so_i = \{e''_{i,j} \mid j \in N\}$ where N is the set of natural numbers. $e''_{i,j} \in so_i$ is defined as $e''_{i,j} = (t''_{i,j}, v_{i,j})$, where $t''_{i,j} \in T''$ and $v_{i,j} \in V''$ giving $so_i \subseteq T'' \times V''$. Events within so_i are totally ordered. j is an index that defines ordering. Given two events $e''_{i,p}$, $e''_{i,q}$, $p < q$ (here $<$ is the ordinary ordering relation for integers) implies $e''_{i,p} < e''_{i,q}$. However, as signals in SO are asynchronous, two events that belong to two different signals $e''_{x,p} \in so_x$ and $e''_{y,q} \in so_y$ are in general not related by $\le$, i.e. $e''_{x,p} \not\le e''_{y,q}$ regardless of p and q.

FI is the set of synchronous input FSM signals that result from the conversion of the SDF output signals. Each $so_i \in SO$ is converted into one $fi_i \in FI$. Thus SO and FI are of the same size m we defined above i.e. $FI = \{fi_i \mid 1 \le i \le m\}$. Each $fi_i \in FI$ is a set of events $fi_i = \{e'_{i,j} \mid j \in N\}$. $e'_{i,j} \in f_i$ is defined as $e'_{i,j} = (t'_{i,j}, vec_{i,j})$

where $t'_{i,j} \in T'$ and $vec_{i,j} \in V''^{n_i} \cup \{\perp\}$, hence $vec_{i,j} = (v''_{i,j,1}, v''_{i,j,2} \ldots v''_{i,j,n_i})$ carries n_i values produced in the SDF domain i.e. $v''_{i,j,1}, v''_{i,j,2} \ldots v''_{i,j,n_i} \in V''$ and it is possible that $vec_{i,j} = (\perp)$ which means that the event is absent. All events within the same signal i have to carry the same number of values n_i. This is due to the rule that the SDF graph must produce the same number of output tokens in each iteration. Events in different signals can carry a different number of values. Within each $fi_i \in FI$, all events are totally ordered as in asynchronous SDFG output signals. Moreover, events are also ordered across all $fi_i \in FI$. Given two events that belong to two different signals $e'_{x,p} \in fi_x$ and $e'_{y,q} \in fi_y$, $p < q$ implies that $e'_{x,p} < e'_{y,q}$. If $p = q$ than $e_{x,p} = e_{y,q}$ which means that $e_{x,p}$ and $e_{y,q}$ are synchronous events. Therefore we can drop the signal index in the tag so that $e'_{i,j} = (t'_j, vec_{i,j})$.

The synchronous signals in *RFI* indicate whether the input FSM channels are ready for rendezvous. In particular, each $rfi_i \in RFI$ shows whether the correspon-ding fi_i is ready to receive data. Data can be sent through fi_i if the rendezvous state that is connected to fi_i is active. Each $rfi_i \in RFI$ is a set of events $rfi_i = \{e'_{rfi_i,j} \mid j \in N\}$, where $e'_{rfi_i,j} = (t'_j, vec_{rfi_i,j})$. The signals in *RFI* are pure. Hence, $vec_{rfi_i,j}$ can take two values. If rfi_i is present in the jth tick, then $vec_{rfi_i,j} = p$, otherwise if it is absent then $vec_{rfi_i,j} = \perp$. The presence of rfi_i indicates that fi_i is ready to receive data, whereas the absence indicates the opposite.

sor (SDFG outputs ready) also belongs to the synchronous domain. It is a set of events $sor = \{e'_{sor,j} \mid j \in N\}$, where $e'_{sor,j} = (t'_j, vec_{sor,j})$ and $vec_{sor,j}$ can take one of the two values, p or $\perp$. What the signals in *RFI* do for the signals in *FI*, *sor* does for the signals in *SO*. If *sor* is present, the output SDFG signals are ready to send data; if it is absent they are not ready. Since SDFG output signals are all ready or not ready at the same time, only one signal is needed to indicate their status.

When *STF* is triggered by a tick of the clock, it invokes the following function for each output channel:

$$stf_i(e'_{sor,j}, e'_{rfi_i,j}, \{e''_{i,k_i+1}, e''_{i,k_i+2}, \ldots, e''_{i,k_i+n_i}\}) = e'_{i,j} \tag{4.1}$$

which is equivalent to

$$\begin{aligned} &stf_i((t'_j, vec_{sor,j}), (t'_j, vec_{rfi_i,j}), \{(t''_{i,k_i+1}, v''_{i,k_i+1}), (t''_{i,k_i+2}, v''_{i,k_i+2}), \ldots, \\ &(t''_{i,k_i+n_i}, v''_{i,k_i+n_i})\}) = (t'_j, vec_{i,j}) \end{aligned} \tag{4.2}$$

with

$$vec_{i,j} = \begin{cases} (v''_{i,k_i+1}, v''_{i,k_i+2} \ldots v''_{i,k_i+n_i}) & \text{if } vec_{sor,j} = p \wedge vec_{rfi_i,j} = p \\ \perp & \text{otherwise} \end{cases}$$

where n_i is a constant that denotes the number of tokens (events in TSM) produced on the channel in one iteration of the SDFG. On the other hand k_i is a variable, which increases with the number of completed iterations. For example, if we assume that five tokens are produced in each iteration of the SDFG, k_i=0 after the first iteration and k_i=5 after the second iteration. If we label the number of completed SDFG iterations as r we can express k_i as

$$k_i = r \cdot n_i \tag{4.3}$$

If $e'_{sor,j}$ and $e'_{rfi,j}$ are present stf_i takes values spread across multiple events in the asynchronous domain and groups them under a single event in the synchronous domain. Otherwise if $e'_{sor,j}$ is absent (SDF graph is not ready to communicate) and/or $e'_{rfi,j}$ is absent (the rendezvous state for fi is not active), then the communication does not occur, so $e'_{i,j}$ is absent.

An important effect of STF is that values carried by asynchronous signals in SO can become synchronous as shown in the example below where there are three channels and five iterations have been completed.

$$stf_1(e'_{sor,j}, e'_{rf_1,j}, \{e''_{1,4\cdot n_1+1}, e''_{1,4\cdot n_1+2}, ..., e''_{1,4\cdot n_1+n_1}\}) = e'_{1,j} = (t'_j, vec_{1,j})$$

$$stf_2(e'_{sor,j}, e'_{rf_2,j}, \{e''_{2,4\cdot n_2+1}, e''_{2,4\cdot n_2+2}, ..., e''_{2,4\cdot n_2+n_2}\}) = e'_{2,j} = (t'_j, vec_{2,j})$$

$$stf_3(e'_{sor,j}, e'_{rf_3,j}, \{e''_{3,4\cdot n_3+1}, e''_{3,4\cdot n_3+2}, ..., e''_{3,4\cdot n_3+n_3}\}) = e'_{3,j} = (t'_j, vec_{3,j})$$

If $e'_{sor,j}$, $e'_{rf_1,j}$, $e'_{rf_2,j}$ and $e'_{rf_3,j}$ are all present in the jth tick, then $e'_{1,j}$, $e'_{2,j}$, $e'_{3,j}$ will also be present and they will carry the values from the asynchronous domain.

4.2.2 Data Transfer from FSM to SDF

FTS performs conversion in the opposite direction. It takes three inputs: the set of signals FO, the set of signals RFO and signal sir. It outputs the set of signals SI. FO is the set of synchronous FSM signals that provide input data for the SDFG. We define that there are m signals in FO so that $FO = \{fo_i \mid 1 \le i \le m\}$. Each $fo_i \subseteq FO$ is a set of events $fo_i = \{e'_{i,j} \mid j \in N\}$. $e'_{i,j} \in fo_i$ is defined as $e'_{i,j} = (t'_j, vec_{i,j})$ where $t'_j \in T'$ and $vec_{i,j} \in V'^{l_i} \cup \{\perp\}$. Thus $fo_i \in T' \times (V'^{l_i} \cup \{\perp\})$ and $vec_{i,j} = (v'_{i,j,1}, v'_{i,j,2} .... v'_{i,j,l_i})$ carries l_i values produced in the synchronous domain i.e. $v'_{i,j,1}, v'_{i,j,2} .... v'_{i,j,l_i} \in V'$. If the event is absent than $vec_{i,j} = (\perp)$. SI is the set of SDF input signals that receive tokens from FSMs. As each $fo_i \subseteq FO$ is converted into one $si_i \in SI$, there are m signals in SI i.e. $SI = \{si_i \mid 1 \le i \le m\}$. Each $si_i \in SI$ is a set of events $si_i = \{e''_{i,j} \mid j \in N\}$. $e''_{i,j} \in si_i$ is defined as $e''_{i,j} = (t''_{i,j}, v_{i,j})$ where $t''_{i,j} \in T''$ and $v_{i,j} \in V'$ giving $si_i \subseteq T'' \times V'$. Ordering of events in synchronous HCFSM and asynchronous SDF signals was already described when STF was discussed.

Each $rfo_i \in RFO$ is a set of events $rfo_i = \{e'_{rfi,j} \mid j \in N\}$, where $e'_{rfi,j} = (t'_j, vec_{rfi,j})$ and $vec_{rfi,j} \in \{\perp, p\}$. If $e'_{rfi,j}$ is present the corresponding rendezvous state is active, if it is absent the state is inactive. On the other hand $sir = \{e'_{sir,j} \mid j \in N\}$ which also belongs to the synchronous domain, shows the status of the SDFG input signals. If $e'_{sir,j}$ is present the SDFG is ready to receive inputs. If $e'_{sir,j}$ is absent it is not.

When *FTS* is triggered by a tick of the clock, it invokes the following function for each output channel:

$$fts_i(e'_{sir,j}, e'_{rf_i,j}, e'_{i,j}) = fts_i((t'_j, vec_{sir,j}), (t'_j, vec_{rf_i,j}), (t'_j, vec_{i,j}))$$
$$= \begin{cases} \{e''_{i,q_i+1}, e''_{i,q_i+2}, \ldots, e''_{i,q_i+l_i}\} \text{ if } vec_{sir,j} = p \wedge vec_{rf_i,j} = p \\ \varnothing \quad \text{otherwise} \end{cases} \tag{4.4}$$

where

$$\{e''_{i,q_i+1}, e''_{i,q_i+2}, \ldots, e''_{i,q_i+l_i}\} = \{(t''_{i,q_i+1}, v'_{i,j,1}), (t''_{i,q_i+2}, v'_{i,j,2}), \ldots, (t''_{i,q_i+l_i}, v'_{i,j,l_i})$$

In (4.4) l_i is a constant that denotes the number of tokens consumed on si_i in one SDF iteration while q_i is a variable which can be expressed as

$$q_i = r \cdot l_i \tag{4.5}$$

where r is the number of completed SDF iterations.

An important effect of *FTS* is that values carried by synchronous signals in *FO* become desynchronised, as shown in the example below, where there are three channels and five iterations have been completed. We also assume that all input events are present.

$$fts_1(e'_{sir,j}, e'_{rf_1,j}, e'_{1,j}) = \{e''_{1,5\cdot l_1+1}, e''_{1,5\cdot l_1+2}, \ldots, e''_{1,5\cdot l_1+l_1}\}$$

$$fts_2(e'_{sir,j}, e'_{rf_2,j}, e'_{2,j}) = \{e''_{2,5\cdot l_2+1}, e''_{2,5\cdot l_2+2}, \ldots, e''_{2,5\cdot l_2+l_2}\}$$

$$fts_3(e'_{sir,j}, e'_{rf_3,j}, e'_{3,j}) = \{e''_{3,5\cdot l_3+1}, e''_{3,5\cdot l_3+2}, \ldots, e''_{3,5\cdot l_3+l_3}\}$$

4.3 The Impact of Clock Speeds

In Sect. 4.1.1, when we discussed determinism we focused on each clock domain separately without paying attention to effects caused by the relation between clocks. The aim of this section is to show how clock speeds can impact the overall behaviour of a system. For some systems, such as the frequency relay, clock speeds are only important for time constraints. For others, such as the one presented in Fig. 4.26, they completely change the system behaviour.

The specification in Fig. 4.26 does not have any input signals, but it has two output signals, *c* and *d*. When FSM1 makes a transition from S3 to S1 both SDF1 and FSM3 are ready to operate. After each iteration, SDF1 outputs integer 3 on ch1 while ch2 is only used for synchronization. FSM3 increments shared variable v2 in every tick.

Both of its transitions are always enabled. If SDF1 always takes three gclk ticks to complete its iteration, the system will not produce any output. If it always takes

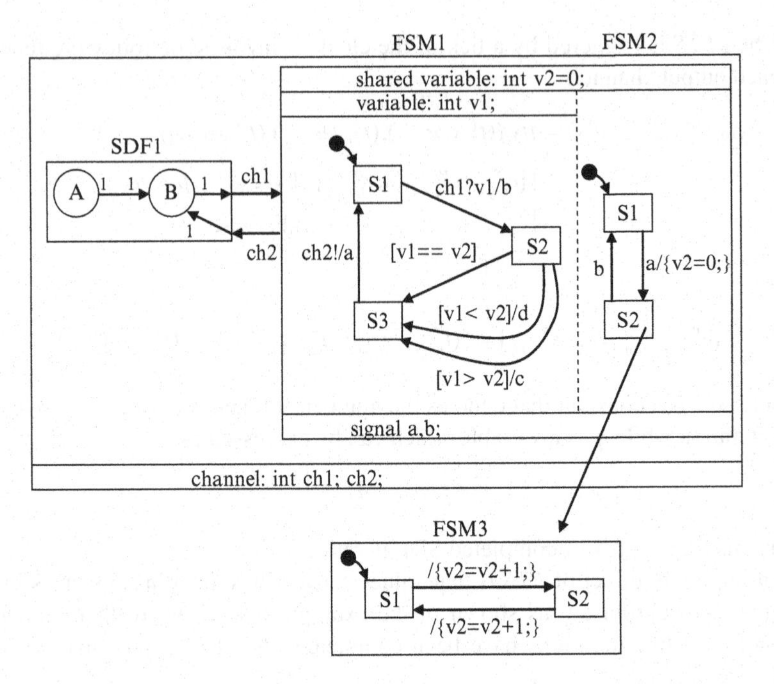

Fig. 4.26 A specification with behaviour sensitive to clock speeds

more than three ticks, a sequence of *c* outputs will be produced, and if it always takes less than three ticks, a sequence of *d* outputs will produced. In addition, any combination of behaviours is possible if the length of SDF1 iterations is not constant in terms of *gclk* ticks.

Determinism is not viewed consistently across the literature. In [94], the authors claim that Multiclock Esterel is deterministic even though the behaviour of a specification can be influenced by clock speeds as in DFCharts. The non-determinism arising from clocks is considered to be external. On the other hand, the Kahn process networks (KPN) model is said to be deterministic exactly because the output sequence is independent of process speeds. Whether we call non-determinism due to clocks external or internal it does pose a problem to design if not handled properly. We could completely avoid this issue in DFCharts semantics by fixing the speed of every SDFG to a constant number of *gclk* ticks. That would be a poor solution, because it would severely reduce implementation space. Instead, we allow experimenting with different speeds of SDFGs relative to *gclk* in DFCharts design flow. As a result, verification becomes more intensive but efficient implementation can be obtained in the end.

Chapter 5
DFCharts in SystemC and Esterel

With graphical syntax presented in Chap. 1 and Java-based textual syntax given in Chap. 6, DFCharts can be used as a language for specification of embedded systems. In this chapter we use DFCharts as a model of computation to assess the effectiveness of two popular system level languages, SystemC and Esterel, in capturing behaviour of heterogeneous embedded systems. While SystemC is being proposed by an industry consortium and has no formal semantics, Esterel has a formal semantics and formal verification capabilities. Hence, both these languages represent differing perspectives on system-level modelling. The frequency relay case study was specified in both languages, following the DFCharts model as closely as possible. Using this analysis, we can identify what needs to be improved in each language in order to increase their ability to handle heterogeneous embedded systems. The relation between the two languages and DFCharts was previously described in [105].

In Sect. 5.1 we list the requirements for SystemC and Esterel according to DFCharts features. For each requirement we discuss the level of support provided by SystemC and Esterel, mainly by using observations made while specifying the frequency relay in both languages. As neither language was able to completely follow the target DFCharts model, we describe the resulting deviations from DFCharts in the frequency relay specifications. In Sect. 5.2, we discuss some numerical results, such as code size and simulation speed, that were obtained after describing the frequency relay in SystemC and Esterel. Finally, Sect. 5.3 suggests some modifications in the semantics and syntax of SystemC and Esterel that would lead to their better support for heterogeneous embedded systems.

5.1 Analysis Based on Requirements

The list of requirements is as follows:

1. Concurrent processes – an essential requirement, which is a precondition for all the other points to follow.

I. Radojevic and Z. Salcic, *Embedded Systems Design Based on Formal Models of Computation*, DOI 10.1007/978-94-007-1594-3_5,

2. Rendezvous communication.
3. Buffered communication between processes and specification of firing rules for dataflow modules.
4. Support for HCFSM model with synchronous communication.
5. Imperative statements to describe data transformations inside SDF actors, as well as smaller computations performed by FSMs.
6. Hierarchy and preemption – multiple processes inside a hierarchical state and termination of lower-level processes instantly upon an exit from the hierarchical state.

5.1.1 Concurrent Processes

5.1.1.1 SystemC

Concurrency is implicitly assumed in SystemC. Processes defined in a single module are concurrent. When multiple modules are connected at any level of hierarchy, they always execute concurrently. In fact, specifying the execution order of modules, such as sequential or pipelined execution available in other languages like SpecC [79], is not possible in SystemC. The designer would have to manipulate the execution order of modules by using control signals.

5.1.1.2 Esterel

Concurrency can be explicitly created using the parallel operator || at any level of hierarchy. The branches in a parallel operator can contain any number of statements. The || operator creates concurrent threads that communicate and synchronize using synchronous broadcast. This is based on the SR model of computation where a global clock is assumed. Inputs and corresponding outputs are generated in the same tick of the global clock, leading to a zero delay model. Also, events generated in any thread are broadcasted to all other threads. Any other form of concurrency will have to be emulated by clever programming.

5.1.2 Rendezvous Communication

5.1.2.1 SystemC

SystemC does not offer any high level construct that implements rendezvous directly. However it should not be difficult to create rendezvous between two processes using *wait* statements.

5.1.2.2 Esterel

As in SystemC, rendezvous cannot be specified directly. It has to be created using a combination of *await* and *emit* statements that need to be programmed appropriately.

5.1.3 Buffers and Firing Rules for Dataflow

5.1.3.1 SystemC

FIFO buffers can be implemented in SystemC by the primitive channel *sc_fifo*. Because of constant data rates, SDF blocks should be implemented as method processes. There should be no need to use less efficient thread processes. However only thread processes that are dynamically scheduled by the SystemC kernel can use sc_fifo buffers. As a result, static scheduling of the SDF model cannot be easily implemented. Moreover, firing rules associated with the SDF actors are not supported by the primitive FIFO channel. The SystemC kernel can activate a thread process as soon as data is available in its input FIFO channel. Hence, the third requirement of buffered communication with static scheduling can not be directly implemented using a SystemC constructs.

Due to the reasons explained above, the three data-dominated blocks were implemented as thread processes that communicate through sc_fifo channels using blocking reads and blocking writes (blocking writes are inevitable as sc_fifo channels must be bounded). Thus the three signal processing blocks appear as a Kahn process network in the SystemC specification rather than an SDF network.

5.1.3.2 Esterel

A FIFO buffer can be implemented as a separate process (C function) in Esterel. In this way computation and communication would be separated, which may be a useful feature if components need to be reused. However, the FIFO process would still be synchronized to the tick signal. Thus the level of abstraction would be lower than in asynchronous SDF buffers. Alternatively, buffers could be implemented as *asynchronous tasks*, but this would lead to integration problems which will be discussed below.

It is clear that Esterel cannot efficiently capture the data-dominated (SDF) part of the frequency relay. Therefore, the SDF blocks performing signal processing (averaging filter, symmetry function and peak detection) must be reactive as all other processes in the system. The event they react to is a sample coming from the analogue-to-digital converter. The problem comes from the fact that all processes must be aligned to a single tick signal, i.e. they have to read inputs and produce outputs at the same time. The most efficient solution for the SDF processes is that the tick signal coincides with the sampling frequency of the AC input signal. On the other hand this would be too slow for the parameter settings block whose inputs may be faster than the sampling frequency, especially in the scenario when it is connected to a high-speed CDMA network. The ticks must be frequent enough to capture all inputs in the system. Thus the rate of the tick signal is determined by the process with the fastest inputs in the system, which is the parameter settings block. The consequence is an implementation which is likely to be inefficient, since the data-dominated blocks have to make their computations faster than they need to. A more efficient implementation would be achieved if the data-dominated blocks were specified as asynchronous tasks taking more than one tick to complete computations. The problem with asynchronous tasks is that they are not handled by Esterel analysis tools. In Esterel Studio, the design environment based around Esterel, every asynchronous task is seen as a black box and has to be simulated in another environment. Return times of asynchronous tasks and values they return have to be entered in the same way as inputs in Esterel Studio simulations. For these reasons, in order to be able to make a complete simulation in Esterel Studio, we had to make the data-dominated blocks behave according to the synchrony hypothesis.

Multiclock Esterel [58] was proposed to alleviate some of the problems described above. In this extension of Esterel, processes may run at different speeds, which implies that there are multiple tick signals in the system. Processes that are triggered by different ticks can use various mechanisms to synchronize, including sampling and reclocking [94].

5.1.4 HCFSM with Synchronous/Reactive Communication

5.1.4.1 SystemC

In SystemC, an FSM can be described with a switch-case construct, which can be nested to describe hierarchical FSMs. This involves using multiple state variables. Figure 5.1 shows a section of the method process that describes the parameter settings and threshold reception FSMs of the frequency relay. The protocol for receiving thresholds starts from state s2_0, which is nested inside state s2. Every time state s2 is entered signals *cancel* and *done* are checked before the execution of the threshold reception protocol. The description in Fig. 5.1 is essentially behavioural level abstraction. It is also cycle-accurate, since the process is driven by a clock.

Simultaneous events, which are necessary for synchronous communication between hierarchical FSMs, can be created in SystemC. However, instantaneous communication is not handled in the same way as in synchronous languages. The SystemC kernel operates according to the discrete event model where simultaneous events are resolved with microsteps. In contrast, a fixed point solution is sought in the semantics of synchronous languages.

Reactivity is supported by signal sensitivities and *wait* statement. In control-dominated systems, a consequence of a reaction is often a pre-emption. SystemC lacks powerful preemption constructs such as *abort* and *trap*. Therefore, it does not fully satisfy the fourth requirement.

```
case s2:
          if (cancel) {
               state = s0;
               substate = s2_0;
          }
          else if (done) {
               state = s3;
               substate = s2_0;
          }
          else
               switch (substate) {
        case s2_0:
                    if (skip) {
                         skip = false;
                         substate = s2_1;
                         thresh_code = 2;
                    }
                    else if....
```

Fig. 5.1 Hierarchical FSMs in SystemC

```
await start;
  loop
  var counter := 0 : integer in
      weak abort
         positive repeat 8000 times
              await tick;
              counter := counter + 1
         end repeat;
         emit time_out;
         halt
      when start
   end var
end loop
```

Fig. 5.2 Esterel specification of timer in frequency relay

5.1.4.2 Esterel

Obviously, the synchronous reactive model is completely supported by Esterel. Esterel contains plenty of statements that enable specifying control-dominated behaviour in a natural way. For example, reactivity is supported by statements such as *await*. Pre-emption can be naturally described by statements such as *abort* and *trap*. This is illustrated in Fig. 5.2, which represents Esterel code for the timer in the frequency relay. When the counting of 8,000 ticks is finished, the block emits signal *time_out*. However, the operation can be preempted by signal *start* causing the counter to reset.

All Esterel specifications, including those that contain pre-emption statements, can be translated into an FSM. The benefit of using pre-emption statements such as

```
while(1) {
        if (measure_off.read()) {
        ..........
        }
        else {
                index = (index+1)%ws;
                window[index] = sample_in.read();
                float sum = 0;
                for (j=0;j<ws;j++)
                        sum = sum + window[(index+j)%ws];
                ave_result.write(sum/ws);  // output average
        }
}
```

Fig. 5.3 Section of SystemC code for averaging filter in frequency relay

abort and *trap* are potentially many states and transitions that are implicit. As a result the specification becomes compact and easily readable. However, there are many situations when a designer wants all states of an FSM to be explicitly represented in the specification. This is where Synccharts, a Statecharts version based on the Esterel semantics, complements Esterel.

5.1.5 Data Transformations

5.1.5.1 SystemC

As an imperative language, SystemC provides an excellent support for describing sequential algorithms and thus it satisfies the fourth requirement quite well. To illustrate this point, a section of the code from the main loop of the averaging filter (a thread process) is shown in Fig. 5.3. The main operation is contained inside the *else branch*. The algorithm reads a sample, performs averaging inside the loop and then the result is written. The control signal *measure_off* is checked at the beginning of the loop. The reason for the presence of this signal will be explained when the next requirement is discussed.

5.1.5.2 Esterel

C is available as a host language in Esterel and, hence, complex algorithms for data transformations inside data-dominated blocks can be specified in a similar way as in Fig. 5.3. As already discussed, the problem is the assumption of instantaneous computation that has to be applied to time-consuming algorithms. Otherwise, they are out of reach of analysis tools in Esterel Studio when modelled as asynchronous tasks.

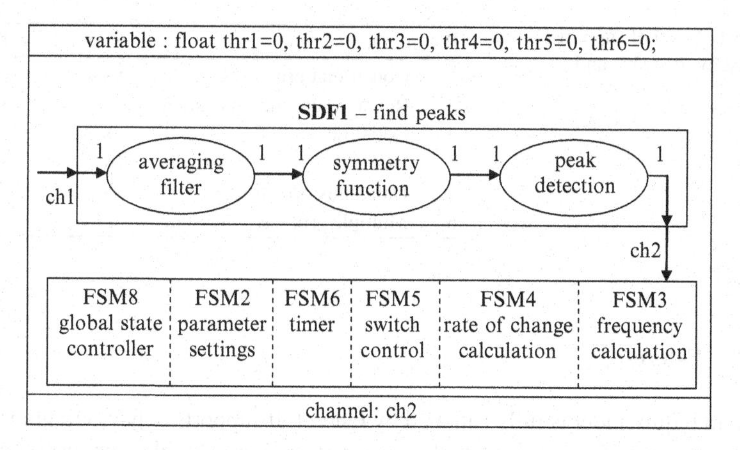

Fig. 5.4 Modified model to better suite the SystemC specification

5.1.6 Multiple Processes Inside a State

5.1.6.1 SystemC

Exception handling is a key component of the frequency relay behaviour as modelled by the top level FSM in Fig. 3.6. When events *off* or *reset* occur, state S2 should be left and all processes in it instantly terminated. SystemC does not fulfil this requirement, since there is no direct way to implement exceptions modelled by exits from higher-level hierarchical states. It was indicated earlier that hierarchy in an FSM could be modelled by nested switch-case statements; this type of modelling is not applicable here since it is not possible to instantiate processes inside a case branch.

As the top level FSM in Fig. 3.6 cannot be directly implemented, the execution of each process has to be controlled by one or more control signals. Due to this, a separate FSM named "global state controller" was added to enable suspension and reactivation of the FSMs and SDFG. It resides with other processes at the same hierarchical level. The control signal *measure_off* in Fig. 5.3 comes from this FSM. If *measure_off* is active, the process does not execute the main operation. It becomes active only when *measure_off* is deactivated.

A modified model of the frequency relay was created to better suit SystemC and this is shown in Fig. 5.4. As already mentioned, the three SDF actors are implemented as threads and cannot be terminated instantaneously. Each of them can change its state from active to inactive, or vice versa, only after reading a control signal, which is done before the beginning of each iteration. This is in contrast to the original model in Fig. 3.6, which assumes the SDFG and all FSMs in state S2 can be terminated instantaneously.

Table 5.1 Level of support provided by SystemC and Esterel

Requirement		SystemC	Esterel
1	Concurrent processes	xxx	xxx
2	Rendezvous communication	xx	xx
3	Dataflow	xx	
4	HCFSM	xx	xxx
5	Data transformations	xxx	xxx
6	Multiple processes in a state		xxx

5.1.6.2 Esterel

The sixth requirement is fully satisfied by Esterel. It supports behavioural hierarchy and has several statements that enable preemption. For example, concurrent blocks can be run inside the abort statement.

5.1.7 Comparison Between SystemC and Esterel

The requirements are listed again in Table 5.1 which also shows the level of support (in a scale of 0–3 as indicated by the number of Xs) for a given feature in SystemC and Esterel. Both languages fully support the requirements one and five. It is also obvious from the previous discussion that SystemC has no support for the sixth requirement while Esterel fully satisfies it. On the other hand, Esterel has no support for dataflow. Although no direct support is available for rendezvous, it can be constructed in both languages. SystemC does not fully satisfy the third requirement since only thread processes can be used with FIFO channels. As a result, only dynamic scheduling is possible. Furthermore, there is no way to directly specify firing rules in SystemC. The fourth requirement is also not fully supported by SystemC due to the lack of pre-emption.

5.2 Numerical Results

The files that comprise the SystemC specification of the frequency relay are shown in Table 5.2 with the number of lines of code for each file. Each file represents a process from the model in Fig. 5.4. The exception is frequency_relay.cpp, the top level file where all lower level processes are connected. In addition, this file also contains the code that specifies the global state controller. At the bottom of the table is the testbench file, which generates input stimuli for the frequency relay and records outputs in a .vcd file.

The Esterel specification was created in Esterel Studio. Table 5.3 shows the list of files and their sizes. This specification is a mixture of Esterel files whose

Table 5.2 SystemC files for frequency relay specification

System C files	Code size (number of lines)
averaging_filter.cpp	85
symmetry_function.cpp	95
peak_detection.cpp	66
frequency_calculation.cpp	93
roc_calculation.cpp	100
parameter_settings.cpp	239
switch_control.cpp	135
timer.cpp	38
frequency_relay.cpp	251
testbench.cpp	412

Table 5.3 Esterel files for frequency relay specification

Esterel files	Code size (number of lines)
Dataflow.strl	76
averaging_filter.c	34
symmetry_function.c	41
measurement.strl	77
freq_average.c	31
roc_average.c	43
parameter_settings.strl	251
switch_control.strl	139
frequency_relay.strl	209

extension is .strl and C files whose extension is .c. In this case, the file structure does not completely follow the model in Fig. 3.6 mainly because of the fact that two languages had to be used. Dataflow.strl represents the three SDFG blocks. It invokes averaging_filter.c and symmetry_function.c, while the peak detection algorithm is described with Esterel statements. Most intensive computations appear inside these two C files, not dataflow.strl. Measurement.strl implements the frequency calculation and rate of change calculation blocks. It uses two C files freq_average.c and roc_average.c for minor computations. Switch_control.strl implements both the switch control and timer blocks. Frequency_relay.strl is the top level file, which connects all lower level blocks but it also implements the top level FSM. As we are only concentrating on syntax for the moment, it is irrelevant whether the C files in Table 5.3 represent instantaneous functions or asynchronous tasks.

The total code size for the SystemC specification excluding the testbench file is 1,102 lines while the total code size for the Esterel specification is 901 lines. This difference is not significant taking into account that the SystemC specification had more files and thus more declarations. It should also be noted that the declarations of C functions used by the Esterel specification are contained in separate Esterel Studio files not listed in the table.

Many different input patterns had to be specified in order to make a satisfactory simulation for the frequency relay. It is no surprise that the testbench is the biggest file in Table 5.2. While both the testbench and system under test can be specified in SystemC, only the actual system can be specified in Esterel. In Esterel Studio, it is possible to run step-by-step interactive simulations. While this feature is useful for quick checks, it would be inefficient for the complete simulation of a large system. Instead all inputs have to be created in advance. For that purpose, Esterel Studio offers *scenario files*. However, writing scenario files manually is tedious, since a very simple language has to be used. It is probably faster to generate a scenario file as an output of a program written in a high level programming language. All this still takes more time than writing a testbench in SystemC.

While the time to prepare a simulation is important, a more important factor to consider is the actual simulation time. It took close to *4 h* for the simulation of the Esterel specification versus just *5 min* for the simulation of the SystemC specification. Both simulations were done on the same computer. The SystemC simulation was done in Microsoft Visual C++ version 6 with SystemC class library 2.0.1. The Esterel simulation was done in Esterel Studio version 4, which supports Esterel v5.

There are probably several factors that caused such huge difference, but the most interesting one is to do with the modelling in Esterel. The whole system has to run on one clock since Esterel does not support multiple clocks. The speed of the system is determined by the process with the fastest changing inputs, which is the parameter setting block. This speed is unnecessarily high for data-dominated parts which need to read inputs only when a sample arrives. As a consequence, there are many ticks with absent inputs in this part of the system.

It should be noted that Esterel model is first converted into C code and then simulated. It is quite possible that conversion into C code is still not as efficient as it could be. This can also have a large impact on the simulation time.

Even though SystemC code can be simulated much faster than Esterel code, we cannot state that in general SystemC has better validation capabilities than Esterel. Although simulation is the most widely used method of validation, it is not the only one. The other method is *formal verification*, which may be applied to Esterel specifications (unlike SystemC). In the frequency relay case study, formal verification has limited use since any useful properties that could be verified would be related to data-dependant internal activities rather than inputs and outputs. It would be difficult to define such properties using Esterel observers that only check properties over the control part.

5.3 Feature Extensions of SystemC and Esterel

According to the analysis presented in Sect. 5.1, SystemC could better support or does not support at all requirements 2, 3, 4 and 6. A rendezvous channel can be constructed by a designer using wait and notify statements to create request and acknowledge lines that are necessary for the rendezvous protocol, although that could take some effort. Ideally, a standard rendezvous channel should be added to the library of channels that includes sc_fifo, sc_signal etc. Asynchronous thread processes that communicate through FIFO channels using blocking reads provide a good foundation for dataflow models. On the other hand, it is still difficult to specify firing rules and construct static scheduling orders. Improvements need to be made in that direction. Synchronous processes can be created in SystemC, which is essential for the fourth requirement. Reactivity can also be modelled using signal sensitivities. On the other hand, the non-existence of preemption is a serious disadvantage when modelling control-dominated behaviour. Processes cannot be instantaneously terminated or interrupted, which is necessary for the sixth requirement. This fundamental limitation can be overcome only by making deep changes in the simulation semantics of SystemC.

SystemC-H [95] is an extension of SystemC, where some desired changes discussed above have already been implemented. The SystemC kernel was extended to provide better support for SDF, CSP and FSM models. In SystemC-H it is possible to construct static schedules for SDF models, which leads to increased simulation efficiency compared to SystemC. Another important addition is hierarchical heterogeneity described in [96], which makes it possible to refine a state of an FSM with an SDFG. In its current form, SystemC-H would probably not be able to support DFCharts entirely since it adheres to purely hierarchical heterogeneity found in Ptolemy. DFCharts represents a mix of hierarchical and parallel heterogeneity.

Like SystemC, Esterel does not directly support rendezvous, but it is possible to construct it using await and emit statements. The main problem of Esterel is a complete lack of support for requirement 3. The assumption made by the synchrony hypothesis that all computations are instantaneous is often not valid for data-dominated systems. Furthermore, Esterel syntax is not appropriate for dataflow. It would be possible to design a dataflow network inside an asynchronous task. However, the asynchronous task is currently just a black box in Esterel's semantics. Describing something in an asynchronous task requires going outside Esterel and its development tools. In order to create a solid basis for an integrated environment, it is necessary to define a model of computation for asynchronous tasks, which could be SDF, and to interface it with the synchronous reactive model.

Chapter 6
Java Environment for DFCharts

This chapter presents a Java environment for DFCharts based design. The specification of finite state machines, synchronous dataflow graphs and communication mechanisms used in DFCharts is supported by a Java class library. Besides the DFCharts library, Ptolemy is needed for simulation of SDF graphs.

A DFCharts specification is created in Java by describing FSMs and SDFGs, and then connecting them. All FSMs and SDFGs used in a specification have to be defined as classes first, and then instantiated at the appropriate place in the specification hierarchy. We will describe FSM and SDFG classes in Sects. 6.1 and 6.2 using examples from the frequency relay specification. Besides FSM and SDFG classes, a DFCharts specification also needs a class that handles the top level of the specification. This type of class will be described in Sect. 6.3. Section 6.4 describes how DFCharts specifications are executed in Java. While the first four sections present the designer's perspective of specification and simulation in Java, Sect. 6.5 describes the class library that enables this process. The chapter is concluded in Sect. 6.6 by looking at differences between the frequency relay specification in Java and those in SystemC and Esterel.

The descriptions in the following sections can also be used as the basis for a DFCharts graphical user interface, which would look exactly like one presented by the figures in Chap. 3. Java provides a wide range of utilities for graphical programming especially in the *swing* package. Graphical objects would need to be converted into instances of FSM and SDFG classes, which should be a straightforward process.

6.1 FSM Classes

Figure 6.1 shows the structure of an FSM class. Each FSM class has to extend *BaseFSM,* a library class. As seen from the figure, the constructor plays a major role in the class definition since it contains several code sections. Each code section will

I. Radojevic and Z. Salcic, *Embedded Systems Design Based on Formal Models of Computation*, DOI 10.1007/978-94-007-1594-3_6,

```
public class FSM extends BaseFSM {

        reference variables for input signals
        reference variables for output signals
        reference variables for shared variables
        reference variables for states
        reference variables for local variables

        inner classes for transition inputs
        inner classes for transition outputs

        FSM (inputs, outputs, shared variables, channels...){      // constructor
                base constructor parameters

                input signal connections
                output signal connections
                shared variable connections
                local variable initializations

                instantiation of states
                instantiation of transition inputs and outputs
                state connections

                instantiation of local signals for lower level FSMs
                instantiation of shared variables for lower level FSMs
                instantiation of channels for lower level FSMs and SDFGs

                instantiation of lower level FSMs
                instantiation of lower level SDFGs

                state refinements
        }
}
```

Fig. 6.1 Structure of FSM classes

```
InputSignal start_roc;
OutputSignal rd;
SharedVariable<Integer> roc_stat;
SharedVariable<Float> ave_freq;
SharedVariable<Float> roc_thresh1;
SharedVariable<Float> roc_thresh2;
SharedVariable<Float> roc_thresh3;
SimpleState s41, s42, s43;
float first, second;   // last two values of ave. freq. used for roc calculation
float temp;
float roc, ave_roc;
int roc_ws;            // buffer size
float[] roc_window;    // buffer containing rate of change samples
```

Fig. 6.2 Signals, variables and states in FSM4

be described below. The modifier *public* is optional. It is not needed if the class does not need to be seen outside its package.

6.1.1 Declaration of Reference Variables for I/O Signals, States and Variables

DFCharts library classes *InputSignal, OutputSignal, HierarchicalState, SimpleState* and, *SharedVariable* represent input signals, output signals, simple states, hierarchical states, and shared variables, respectively. Figure 6.2 shows the part of class FSM4 (rate of change calculation) which declares variables for these data types.

In the frequency relay specification in Sect. 3.2, *rd* and *start_roc* appear as local signals, but from the point of view of FSM4, they are seen as input and output signals. Their role as local signals becomes apparent when FSM1 is constructed. This detail will be shown later in the section. While input and output signals are pure in DFCharts, shared variables are associated with data types. Currently only primitive data types are supported such as float, integer, and boolean. They have to be used in the form of wrappers. For example, Float is written instead of float. On the other hand, local variables can be of any data type. The local variables of FSM4, which are declared after the states, all have primitive data types.

6.1.2 Inner Classes for Transition Inputs and Transition Outputs

In the general transition form *t[c]/O,P*, the part before the slash is called *transition input* while the part after the slash is called *transition output*. A transition input consists of signals and variable conditions. In Java, it is defined as a class that

Fig. 6.3 The input for transition S41 → S42 in FSM4

```
class s41input1 extends TransitionInput {
   public boolean evaluateInput( ) {
      if (start_roc.read( ))
         return true;
      else
         return false;
   }
}
```

Fig. 6.4 The output for transition S43 → S41 in FSM4

```
class s43output1 extends TransitionOutput {
   public void computeOutput() {
      rd.write(true);
   }
}
```

Fig. 6.5 The output for transition S42 → S43 in FSM4

```
class s42output1 extends TransitionOutput {
   public void computeOutput() {
      float rt1, rt2, rt3;
      rt1 = roc_thresh1.read();
      rt2 = roc_thresh2.read();
      rt3 = roc_thresh3.read();
      if (ave_roc < rt1)
         roc_stat.write(0);
      else if (ave_roc >= rt1 && ave_roc < rt2)
         roc_stat.write(1);
      else if (ave_roc >= rt2 && ave_roc < rt3)
         roc_stat.write(2);
      else if (ave_roc >= rt3)
         roc_stat.write(3);
   }
}
```

extends the DFCharts library class called *TransitionInput. TransitionInput* is an abstract class that contains the single abstract method called evaluateInput(). Thus, a class that extends *TransitionInput* has to have a method that implements evaluateInput(). Figure 6.3 shows the class that specifies the input for the transition that leads from S41 to S42 in FSM4.

InputSignal has the read() method which returns true if the signal is present. The return type of evaluateInput() is boolean, so either true or false must be returned.

A transition output can contain emitted signals and procedures. It is defined as a class that extends the DFCharts library class called *TransitionOutput. TransitionOutput* is an abstract class with the single abstract method called computeOutput(). All classes that extend it must implement this method. Figure 6.4 shows the class that specifies the output for the transition that leads from S43 to S41 in FSM4. Another example is given in Fig. 6.5, which is the output of the transition from S42 to S43.

Emitting output signals and computing procedures are both done inside computeOutput() method. In Fig. 6.4, just one output is emitted by passing true to write() method of OutputSignal. The use of SharedVariable class is illustrated in Fig. 6.5.

```
FSM4 (BaseFSM upper,
      String nameStr,
      InputSignal start_roc_in,
      OutputSignal rd_out,
      SharedVariable<Integer> roc_stat_in,
      SharedVariable<Float> ave_freq_in,
      SharedVariable<Float> roc_thresh1_in,
      SharedVariable<Float> roc_thresh2_in,
      SharedVariable<Float> roc_thresh3_in)

super(upper, nameStr, 3, rd_out);
```

Fig. 6.6 Parameters of FSM4 constructor

It can be noticed that the value of a SharedVariable object is read using read() method and written using write() method.

It may seem unnecessary that transition inputs and outputs have to be specified as classes. It would be easier to simply create methods that are not inside classes. However, *BaseFSM* class in the DFCharts library, from which all FSMs have to inherit, would not be able to handle execution in that case. The reason is the fact that a method in Java can only be invoked with its name. It is not possible to create a variable that points to a method. On the other hand, BaseFSM is able to call transition inputs and outputs, which implement abstract classes.

6.1.3 Constructor Parameters

The parameters passed to the constructor of an FSM include references for input signals, output signals, shared variables and internal channels. In addition, the name is needed as well as the reference for the refined (parent) FSM. It is also possible for an FSM to be instantiated at the top level. In that case the reference for the top level class is passed to the constructor.

The base constructor has to be handled as well. It takes the reference to the object that instantiates the FSM (another FSM or top level), the FSM name, the number of states, and references to output signals.

All input and output signals are created at the top level. A reference to a SharedVariable object may be passed to an FSM, but an FSM can also create a shared variable inside its constructor, if it is needed for lower level FSMs. Figure 6.6 shows the parameters for the constructor of FSM4.

6.1.4 Signal and Shared Variable Connections, Initialization of Local Variables

References to signals and shared variables that are passed to a constructor need to be assigned to variables in an FSM class. Figure 6.7 shows how this is done for FSM4. The initialization of local variables can also be seen in the figure.

Fig. 6.7 Connections and initializations in FSM4

```
start_roc = start_roc_in;
rd = rd_out;
roc_stat = roc_stat_in;
ave_freq = ave_freq_in;
roc_thresh1 = roc_thresh1_in;
roc_thresh2 = roc_thresh2_in;
roc_thresh3 = roc_thresh3_in;
first = 50;
second = 50;
roc_ws = 4;
roc_window = new float[roc_ws];
for(int i = 0;i < roc_ws;i++)
   roc_window[i] = 0;
```

Fig. 6.8 Creating transitions in FSM4

```
s1 = new SimpleState(this, 1, "s1");
s2 = new SimpleState(this, 1, "s2");
s3 = new SimpleState(this, 1, "s3");
s1input1 s1t1 = new s1input1();
s2input1 s2t1 = new s2input1();
s1output1 s1o1 = new s1output1();
s2output1 s2o1 = new s2output1();
s3output1 s3o1 = new s3output1();
connect(s1, s1t1, s1o1, s2, 1);
connect(s2, s2t1, s2o1, s3, 1);
connect(s3, s2t1, s3o1, s1, 1);
s1.setInitial();
```

6.1.5 *Linking States, Transition Inputs and Transition Outputs*

In this part of the constructor, objects for states, transition inputs and transition outputs are created and then linked. Figure 6.8 shows this for FSM4. The parameters of SimpleState constructor are the reference to the FSM the state belongs to, the number of outgoing transitions and the state name. HierarchicalState constructor has the same parameters. States, transition inputs and transition outputs are linked to make complete transitions by the method inherited from BaseFSM called connect(). The parameters of this method are source state, transition input, transition output, sink state and transition priority. The last statement in Fig. 6.8 sets S41 as the initial state.

The remaining sections of the constructor are only present in FSMs that are refined by other FSMs or SDFGs. Thus, we have to depart from FSM4. We will use FSM1 from the frequency relay specification for illustrations.

6.1.6 *Local Signals, Shared Variables and Channels for Lower Level FSMs and SDFGs*

Before lower level FSMs and SDFGs can be instantiated, the objects that enable communication between them (local signals, shared variables and internal channels) have to be created. Showing all local signals, shared variables and channels that are created

```
InputSignal start_roc_in = new InputSignal("start_roc_in");
OutputSignal start_roc_out = new OutputSignal("start_roc_out");
LocalSignal start_roc_local = new LocalSignal("start_roc_local",this);
start_roc_in.setLocal(start_roc_local);
start_roc_out.setLocal(start_roc_local);

SharedVariable<Float> ave_freq = new SharedVariable<Float>("ave_freq",0.0f,s12);

SDFFSMIntegerChannel ch2 = new SDFFSMIntegerChannel ("ch2");
```

Fig. 6.9 Signal start_roc, shared variable ave_freq and channel ch1 in FSM1

within FSM1 would take too much space. Thus, we will only show, in Fig. 6.9, local signal *start_roc* that is used by FSM3 and FSM4, shared variable *ave_freq* that is also used by FSM3 and FSM4, and internal channel *ch2* that connects FSM3 and SDF1.

6.1.6.1 Local Signal

A local signal is created as an instance of class LocalSignal. As we mentioned before, lower level FSMs are never aware that they communicate through a local signal. They see it either as an input signal or output signal. For that reason, besides an instance of LocalSignal, instances of InputSignal and OutputSignal also need to be created and passed to appropriate FSMs. Then, they are linked with the instance of LocalSignal. For that purpose, both InputSignal and OutputSignal contain method setLocal(), which accepts the reference to a local signal. In Fig. 6.9, *start_roc_out* and *start_roc_in* are passed to FSM3 and FSM4, respectively, and linked with *start_roc_local.*

Only a string denoting the signal name has to be passed to the constructors of InputSignal and OutputSignal. The constructor of LocalSignal also requires the reference to the object that instantiates it, which can be either an FSM or top level specification.

In many applications, local signals are used for communication between FSMs that are on different hierarchical levels. For the higher level FSM, a local signal is always seen as an input signal. The reference for the InputSignal object has to be declared outside the constructor. Suppose that in the frequency relay specification FSM1 also has to read *start_roc* besides FSM4. This situation would be handled in exactly the same way as in Fig. 6.9, except that *start_roc_in* would be declared outside the constructor in the area where all input and output signals for FSM1 are declared. In this way, *start_roc_in* is visible to FSM1 transition inputs.

6.1.6.2 Shared Variable

Creating a shared variable is easier. There are three parameters that have to be passed to the constructor of SharedVariable: the name of the shared variable, the initial value, and the hierarchical state in which the shared variable is active. This is necessary to indicate when the shared variable has to be reinitialized. If a shared

Fig. 6.10 The input for transition S31 → S32 in FSM3

```
class s1input1 extends TransitionInput {
  public boolean evaluateInput() {
    if (ch1.receive(din))
      return true;
    else
      return false;
  }
}
```

variable is created at the top level then it is always active. Instead of the reference to a hierarchical state, the third parameter of the constructor would be the reference to the top level class.

It can be seen in Fig. 6.9 that the initial value of *ave_freq* (average frequency) is 0, and it is active in state S12. Whenever FSM1 enters S12, *ave_freq* is initialized to 0. If FSM1 also needed to access *ave_freq*, this shared variable would not be created in the constructor of FSM1. It would have to be created at the level above FSM1, which is the top level in this case.

6.1.6.3 Channel

When an FSM class obtains the reference to a shared variable, it is allowed to read and write the shared variable. It is up to the designer to ensure that in any tick only one FSM can write. If multiple updates occur, the result becomes unpredictable.

Currently, the DFCharts library supports passing arrays of two data types, integer and double, between FSMs and SDFGs, in both directions. Thus, there are four different types of channels: FSMSDFIntegerChannel, SDFFSMIntegerChannel, FSMSDFDoubleChannel, SDFFSMDoubleChannel. The constructors of all four classes take only the channel's name. All four classes have receive() and send() methods whose parameter is either integer or double array. In Fig. 6.9, an SDFFSMIntegerChannel object is instantiated to enable communication between FSM3 and SDF1. It appears in the input of the transition that leads from S31 to S32 in FSM3. The inner class that specifies the transition input is shown in Fig. 6.10.

Variable *din* is a reference to an integer array. Its single element contains the number of samples between two consecutive peaks in the AC waveform. receive() and send() methods return either true or false depending on whether the rendezvous is happening. In Fig. 6.10, if the rendezvous is happening on ch1 in the current tick, receive() puts data in *din* returns true. If not, receive() returns false while *din* remains unchanged.

6.1.7 Instantiation of Lower Level FSMs and SDFGs

When FSM and SDFG objects are instantiated, all required parameters have to be passed to the constructors in the correct order. In HDL terminology, this would be positional association. Named association is not available. Figures 6.11, 6.12 and

```
FSM3 fsm3 = new FSM3 (this, "fsm3", rd_in, start_roc_out,
                       freq_stat, ave_freq,
                       freq_thresh1, freq_thresh2, freq_thresh3,
                       ch2);
```

Fig. 6.11 Instantiation of FSM3

```
FSM4 fsm4 = new FSM4 (this, "fsm4", start_roc_in, rd_out,
                       roc_stat, ave_freq,
                       roc_thresh1, roc_thresh2, roc_thresh3);
```

Fig. 6.12 Instantiation of FSM4

```
int[] k = {5};
SDFPeriodCalculation sdfpc = new SDFPeriodCalculation (this, "sdfpc", 1, ch2,k);
ch1.setGraph(sdfpc);
```

Fig. 6.13 Instantiation of SDF1

Fig. 6.14 Refinement of S2 in FSM1

```
BaseFSM s1fsmr[] = {fsm2, fsm3, fsm4, fsm5, fsm6};
BaseSDF s1sdfr[] = {sdfpc};
s12.refine(s1fsmr,s1sdfr);
```

6.13 show the instantiation of FSM3, FSM4 and SDF1 in the constructor of FSM1, respectively. The parameters of SDFG constructors will be described in Sect. 6.2. After an SDFG has been instantiated, its channels have to be linked with it using method setGraph() as shown in Fig. 6.13.

6.1.8 State Refinement

After FSMs and SDFGs have been instantiated, the states they refine have to be specified. Those that refine the same state are executed concurrently. HierarchicalState class has method refine(), which is overloaded three times to support all three types of state refinement in DFCharts. It can accept an array of BaseFSM objects, an array of BaseSDF objects, or both at the same time. Figure 6.14 shows the refinement of state S12 in FSM1.

6.2 SDFG Classes

Each SDFG class has to extend the DFCharts library class called BaseSDF. The structure of an SDFG class is shown in Fig. 6.15.

As an example, we use the code for SDF1 listed in Fig. 6.16.

```
public class SDF extends BaseSDF {

        SDF (internal channels, iteration length...) {        // constructor
                base constructor parameters
                try {
                        instantiation of SDF actors

                        actor connections

                } catch {Exception e) {
                        e.printStackTrace();
                }
        }
}
```

Fig. 6.15 Structure of SDF classes

```
public class SDF1 extends BaseSDF {

  SDF1 (BaseFSM container, String graphName,
     int channum, SDFFSMIntegerChannel ch1_in,
     int[] iterationLength) {
   super(container, graphName, channum, iterationLength);

   try {
     TestSamples ts = new TestSamples(top,"ts");      // sample source
     AverageFilter ave = new AverageFilter(top,"ave",20); // averaging filter
     SymmetryDetection sym = new
     SymmetryDetection(top,"sym",80);   // symmetry detection
     PeakDetector pk = new PeakDetector(top,"pk");  // peak detector
     int[] intOutInit = {-1};    // initial token value
     OutputIntegerActor intOut = new
        OutputIntegerActor(top,"IntOut",1,intOutInit); // interface actor
     ch1_in.setActor(intOut);

     top.connect(ts.output,ave.input);
     top.connect(ave.output,sym.input);
     top.connect(sym.output,pk.input);
     top.connect(pk.output,intOut.input);
   } catch (Exception e) {
     e.printStackTrace();
   }
  }
}
```

Fig. 6.16 Class that specifies SDF1

6.2.1 Constructor Parameters

The parameters of SDFG constructors are: the reference to the object that does the instantiation (FSM or top level), SDFG name, references to internal channels, and an integer array that indicates the length of SDF iterations in terms of FSM ticks. An array allows a repeating pattern of SDF iteration lengths. If all iterations have the same length, then the array only needs to have a single element as in Fig. 6.13. The parameters that have to be passed to the constructor of BaseSDF are: the reference to the object that does the instantiation, SDFG name, the number of internal channels and the iteration length array.

6.2.2 Instantiation of Actors

Inside the try-catch statement, SDFG actors are instantiated and connected following the rules of Ptolemy. When actors are constructed, various exceptions that can be thrown have to be caught. Actor constructors can have various parameters. A compulsory parameter that always has to be included is *top*, an instance of Ptolemy class TypedCopmositeActor that is created in BaseSDF. An SDFG in a DFCharts specification is in fact seen as a TypedCompositeActor in Ptolemy.

Design of Ptolemy actors is described in [97]. We briefly illustrate it here using an example in Fig. 6.17., which lists the code for the averaging filter in SDF1. AverageFilter extends Ptolemy class Transformer, which provides utilities for input-output operations. Variables *input* and *output* that appear in the constructor are inherited from this class. Their data type is set to DOUBLE, which is a replacement for the primitive type double in Ptolemy.

During the execution of a Ptolemy actor, methods preinitialize(), initialize(), prefire(), fire(), postfire() and wrapup() are invoked. These methods are not compulsory and most of actors in Ptolemy libraries do not have all of them. However, most have fire(), which is supposed to contain the main functionality of an actor. In AverageFilter, initialize() sets all samples in the buffer (AveWindow) to zero while fire() computes the average of all samples that are in the buffer.

SDFGs that communicate with FSMs, such as SDF1, have to include special interface actors. There are four interface actors corresponding to four types of channels: InputIntegerActor, InputDoubleActor, OutputIntegerActor and OutputDoubleActor. InputIntegerActor matches FSMSDFIntegerChannel, for example. Besides *top*, the constructor of an interface actor is passed the actor name, an integer showing the number of tokens communicated on the channel in a single rendezvous, and an array with initial token values if tokens flow from SDF to FSM. In Fig. 6.17, the peak detection actor sends data to FSM3 by placing it in the interface actor *intOut*. When an interface actor is instantiated, the corresponding channel has to be linked with it using method setActor() as shown in Fig. 6.16.

Sink and source actors, which are needed for testing, have to be attached to external channels. The simple input/output facility described in Sect. 6.4 only handles

```
public class AverageFilter extends Transformer {
  int AveIndex;
  int AveWs;
  double AveWindow[];

  public AverageFilter (CompositeEntity container, String name, int AweWsIn)
        throws NameDuplicationException, IllegalActionException {
    super (container, name);
     AveIndex = -1;
     AveWs = AweWsIn;
     AveWindow = newdouble[AveWs];
     input.setTypeEquals(BaseType.DOUBLE);
     output.setTypeEquals(BaseType.DOUBLE);
  }

  public void initialize() {
    for(int j = 0;j < AveWs;j++)
        AveWindow[j] = 0;
  }

  public void fire() throws IllegalActionException {
    double sample;
    sample = ((DoubleToken) input.get(0)).doubleValue();
    AveIndex = (AveIndex+1) % AveWs;
    AveWindow[AveIndex] = sample;
    double sum = 0;
    for (int j=0;j<AveWs;j++)
    sum = sum + AveWindow[(AveIndex+j) % AveWs];
  output.send(0, new DoubleToken(sum/AveWs));
 }
}
```

Fig. 6.17 Averaging filter actor in SDF1

FSM inputs and outputs. There are no sink and source actors in the DFCharts library since Ptolemy has plenty. It should be noted that the source actor used in SDF1 (class TestSamples) was not taken from a Ptolemy library.

6.2.3 Actor Connections

The actors that make up a TypedCompositeActor are connected with its method connect(). While connecting outputs to inputs, it has to be ensured that data types are compatible [97].

6.3 Top Level Classes

A top level class is needed to instantiate and connect top level FSMs and SDFGs. It has to extend the DFCharts library class called DFChartsTop, which is in many ways similar to HierarchicalState. The structure of a top level class is shown in Fig. 6.18.

The top level class for the frequency relay is shown in Fig. 6.19.

```
public class TopLevel extends DFChartsTop {

    TopLevel (input file, output file) {        // constructor
        base constructor parameters

        instantiation of input signals
        instantiation of output signals

        instantiation of local signals for top level FSMs
        instantiation of shared variables for top level FSMs
        instantiation of channels for top level FSMs and SDFGs

        instantiation of top level FSMs
        instantiation of top level SDFGs

        top level refinement
    }
}
```

Fig. 6.18 Structure of top level classes

6.3.1 *Constructor Parameters*

The constructors of top level classes only have two parameters, the reference for the file that contains input stimulus, and the reference for the file where the outputs will be printed. The references for the two files are also passed to the base constructor.

6.3.2 *Instantiation of Input and Output Signals*

All input and output signals, except those that serve as local signals, have to be instantiated at the top level and passed to FSMs. The constructors of InputSignal and OutputSignal only need the signal name and the reference to the top level class.

6.3.3 *Local Signals, Shared Variables and Channels for Top Level FSMs and SDFGs*

This section is the same as in FSM classes. The top level class of the frequency relay does not have it, since there is only a single FSM at the top.

Fig. 6.19 Top level of frequency relay

```
public class FrequencyRelayTop extends DFChartsTop {

  FrequencyRelayTop (String infName, String outfName) {
    super(infName, outfName);

    InputSignal on = new InputSignal("on",this);
    InputSignal off = new InputSignal("off",this);
    InputSignal reset = new InputSignal("reset",this);
    InputSignal sth = new InputSignal("sth",this);
    InputSignal cancel = new InputSignal("cancel",this);
    InputSignal done = new InputSignal("done",this);
    InputSignal thresh0 = new InputSignal("thresh0",this);
    InputSignal thresh1 = new InputSignal("thresh1",this);
    InputSignal skip = new InputSignal("skip",this);
    OutputSignal nt = new OutputSignal("nt", this);
    OutputSignal inth = new OutputSignal ("inth", this);
    OutputSignal sw1 = new OutputSignal("sw1",this);
    OutputSignal sw2 = new OutputSignal("sw2",this);
    OutputSignal sw3 = new OutputSignal("sw3",this);

    FSM1 fsm1 = new FSM1 (this, "fsm1",
        on, off, reset,
        sth, cancel, done, thresh0, thresh1, skip, nt, inth,
        sw1, sw2, sw3);

    BaseFSM tr[] = {fsm1};
    topLevel(tr);
  }
}
```

6.3.4 Instantiation of Top Level FSMs and SDFGs

This section is also the same as in FSM classes. In the frequency relay, FSM1 accepts the references for all input and output signals even though it only uses *on, off* and reset. The rest is passed to lower level FSMs.

6.3.5 Top Level Refinement

The top level can be seen as always active hierarchical state that does not belong to any FSM. Hence, it is "refined". The topLevel() method in DFChartsTop is very similar to refine() in Hierarrchical state. It is overloaded three times to support the three refinement types. FSM and SDF objects have to be put into arrays before they are passed to topLevel().

6.4 Simulation

Finally, we need a class that runs the top level class. The top level class is executed with runTopLevel(). The class that runs the frequency relay top level is shown in Fig. 6.20.

The input and output files passed to the top level class only contain FSM inputs signals and outputs signals. As mentioned before, SDFGs have separate source and sink actors for testing.

In the input file, the status of each input signal has to be defined in each tick. The name of an input signal and its status (present or absent) are written on separate lines. Ticks are separated by blank lines. If the same inputs repeat over successive ticks, the instruction 'repeat' can be used with the number of ticks that have identical inputs. The simulation is terminated by writing 'end'. An input file for a system with two inputs could look as in Fig. 6.21

In the output file, outputs are printed in the format that is used for inputs. In addition, the current state of each FSM is printed in every tick.

6.5 Library Classes

DFCharts library classes enable execution of FSM and SDF classes, described in the previous sections, by providing synchronization and communication mechanisms. It should be noted that the library does not yet include any classes that identify

```
public class FrequencyRelay {
   public static void main (String args[]) {
      FrequencyRelayTop relay = new
      FrequencyRelayTop("frequency_relay_input.txt","frequency_relay_output.txt");
      relay.runTopLevel();
   }
}
```

Fig. 6.20 Execution of the top level class of the frequency relay

```
input1
present
input 2
absent

repeat
10

end
```

Fig. 6.21 Input file format

causality cycles in DFCharts specifications. Thus, it is currently the responsibility of a designer to ensure that communication between FSMs is valid. If FSMs are incorrectly connected with instantaneous loops, a simulation will deadlock. On the other hand, if any of SDF graphs in a specification is not constructed correctly, Ptolemy software will throw an exception.

In total, there are 23 classes in DFCharts library. They can be divided in five groups: base classes, FSM component classes, FSM communication classes, FSM – SDF communication classes, and synchronization class. The base classes include BaseFSM, BaseSDF, DFChartsTop; the FSM component classes include SimpleState, HierarchicalState, TransitionInput, TransitionOutput, Transition; FSM communication classes include InputSignal, OutputSignal, LocalSignal, SharedVariable; FSM–SDF communication classes include FSMSDFChannel, SDFFSMChannel, FSMSDFIntegerChannel, SDFFSMIntegerChannel, FSMSDFDoubleChannel, SDFFSMDoubleChannel, InputIntegerActor, OutputIntegerActor, InputDoubleActor, OutputDoubleActor. The single synchronization class is ThreadControl.

Most of these classes have already been described in the previous sections to some extent. In the following sections, further details will be added for each group. In addition, relations among different groups will be highlighted.

6.5.1 Base Classes

FSM and SDF objects, which inherit from BaseFSM and BaseSDF, need to be able to run as separate threads. For this reason, both BaseFSM and BaseSDF implement the Runnable interface. A DFCharts top object is attached to the main thread. At any point in the execution of a DFCharts specification in Java, the number of active threads is equal to the sum of all active non-refined FSMs and all active SDFGs, plus the main thread. Active FSMs that are refined do not consume additional threads. The thread of a refined active FSM simply takes over one of the refining FSMs.

BaseFSM provides connect() method for construction of an FSM, which links states with transitions. It also has variables for current state and next state, which are necessary for execution of an FSM.

The main purpose of BaseSDF is to provide an interface with Ptolemy software, which is necessary for execution of SDF graphs. Therefore, when DFCharts library is built, Ptolemy must be included in the build path. From the Ptolemy's perspective, each SDF graph is an instance of TypedCompositeActor, which is called *top* in the DFCharts library. The execution of a TypedCompositeActor is handled by methods provided by class Manager. Those methods are invoked from the run() method of BaseSDF. When an SDFG is started, initialize() is invoked. For each iteration, iterate() is invoked. Finally, when an SDFG is stopped, wrapup() is invoked.

A DFChartsTop object spawns the threads for top level FSMs and SDFGs when runTopLevel() calls one of the three methods – runType1(), runType2() or runType3(), depending on the contents of the top level.

6.5.2 *FSM Component Classes*

All FSM component classes except Transition represent inputs for the connect() method of BaseFSM. The Transition class is used in the data structure that is built inside BaseFSM. It contains reference variables for TransitionInput and TransitionOutput.

The most important function of SimpleState is to determine which transition should be taken and produce the outputs for the selected transition. For that purpose, the method evaluateTransitions() is used. It evaluates transitions inputs in the order of their priorities. When it finds a transition that is enabled, it produces the outputs for that transition. The next state cannot be immediately set since it may be possible that the FSM is pre-empted in the current tick. Consequently, the evaluateTransitions() method of the refined state is called. This is a recursive process that leads to a top level FSM. When it is finished, the next state variable in all surviving FSMs can be updated.

HierarchicalState also has to evaluate transitions, but it has an important additional task. It has to spawn threads for the refining FSMs and SDFs. In this respect, it is very similar to DFChartsTop. If a state is refined only by FSMs, the method runType1() is used. If it is refined by an SDFG, runType2() is used. It should be noted that a state can also be refined by several disconnected, independently operating SDFGs, but this is not a usual situation. If a state is refined by both FSMs and SDFGs, runType3() is used.

TransitionInput and TransitionOutput are abstract classes with no functionality, which contain abstract methods evaluateInput() and computeOutput(), respectively. Their purpose is to facilitate building the data structure that represents an FSM.

6.5.3 *FSM Communication Classes*

InputSignal and OutputSignal enable communication between FSMs and the external environment. Both are relatively simple as they only contain two methods, read() and write(). If the status of a signal is present, boolean true is written; otherwise, boolean false is written.

LocalSignal is more complicated as it has three values: present, absent and unresolved. When a new tick begins, the value of a local signal remains unresolved until an FSM writes true or false. If another FSM attempts to read the local signal while it is still unresolved, its thread will become blocked. It will be notified when the value of the signal becomes true or false.

SharedVariable is simpler than LocalSignal, but more complicated than InputSignal and OutputSignal. It has write() and read() methods which work with all primitive data types. It also contains a flag that indicates when a shared variable is active. A shared variable is reset to its initial value only in ticks in which it gets activated. By comparison, LocalSignal does not need a similar flag, since it does not have memory. At the beginning of each tick, all local signals are set to unresolved.

6.5.4 FSM-SDF Communication Classes

In this group FSMSDFChannel and SDFFSMChannel are base classes. FSMSDFIntegerChannel and FSMSDFDoubleChannel extend FSMSDFChannel, while SDFFSMIntegerChannel and SDFFSMDoubleChannel extend SDFFSMChannel, Channel classes enable communication between FSMs and SDFGs in both directions using arrays of two primitive data types, floats and integers. Arrays always have to be used. If there is only a single value flowing through a channel, an array of size one is used. Communication is performed by receive() and send() methods. The parameter of both methods is the reference to an array, which is read from or written to depending on which method is used. Both methods return true if rendezvous occurs, or false otherwise. Channel classes also contain control variables, which ensure that data flows through a channel only once in a single rendezvous between an FSM and an SDFG.

Input interface actors are essentially source actors in an SDGF when communication is performed with an FSM. They are included in the iteration schedule of an SDFG, just like other actors. The send() method of the channel classes places tokens (floats or integers contained in an array) into an input interface actor. During an iteration of an SDFG, these tokens are read out. It has to be ensured that as many tokens are placed as needed. Otherwise, the input interface actor will output some values multiple times, which would most likely lead to incorrect behaviour.

Similarly, output interface actors can be thought of as sink actors in an SDFG. During an iteration, tokens are placed inside an output interface actor. They are brought to an FSM by the receive() method of the channel classes. As in the case of input interface actors, it needs to be ensured that the right number of tokens is placed in output interface actors.

6.5.5 Synchronization Class

The single class that falls under this category is called ThreadControl. Synchronizing threads is its main purpose, but it additionally performs several other actions that occur at the end of each tick. Among those actions is ensuring that each SDF iteration lasts the specified number of ticks.

When an FSM completes a transition, its thread blocks as it encounters wait(). When all FSM threads block, the tick is completed. At this point, methods in ThreadControl print output signals, read inputs for the next tick, clear local signals and update shared variables. Then, if there are any SDFGs that are due to complete an iteration, ThreadControl waits before starting a new tick. A new tick is started by unblocking all threads using notify().

6.6 Frequency Relay Revisited

When compared against its counterparts in SystemC and Esterel, the most important feature of the frequency relay specification in Java is that it fully conforms to the DFCharts model from Sect. 3.2. The specifications in SystemC and Esterel contain alternative solutions and workarounds for the parts of the DFCharts model that the two languages could not completely support.

The size of the Java specification is 1,479 lines. This appears to be significantly longer than 1,102 lines of SystemC code and 901 lines of Esterel code. However, the difference is entirely due to numerous class declarations that have to be made in Java. As we explained in Sect. 6.1, inputs and outputs of transitions have to be specified as classes. What really matters is that describing the actual behaviour does not take more effort in Java than in SystemC and Esterel. Class declarations can be easily handled with templates. Hence, they should have no impact on design time.

Chapter 7
Heterogeneous Reactive Architectures of Embedded Systems

7.1 Background and Trends

Long-term trend in semiconductor development has been integration of larger and larger systems on a single chip. It has been governed by Moore's law in terms of integration capabilities and double increase of the number of transistors on a chip every 12–24 months. At the same time this has also led to the increasing raw computation power on a single chip. However, all those transistors can't be used to achieve faster and more powerful processors due to architectural and power limitations. Rather, the development has taken another direction towards systems on chip which consists of many processors or processing elements, sometimes tens or hundreds of such elements, with a clear trend towards chips which will have thousands of processors. One of the main reasons for this trend is that those processors are simpler, easier to implement and work at lower frequencies than high performance processors, thus they are less power and energy demanding. However, the new approach, which is often referred to as multiple processor (or multiprocessor) systems on chip (MPSoC) faces many challenges. Among them most notable are related to the selection of the type of processing elements (e.g. general purpose vs. application-specific, uniform vs. heterogeneous), selection of interconnect structures and system architecture (e.g. networks on chip vs. circuit interconnect vs. buses), life-time of the processing elements (static or dynamic or reconfigurable), run-time support (operating systems or customized), design flow and tools support (e.g. traditional programming languages vs. concurrent languages). Many new architectures have emerged with a claim of their advantages over others in at least specific application domains. The new approaches are mostly based on concentration on certain features (e.g. architecture, run-time support or languages) but not many of them look at the big picture and design flow that will ensure more consistency and better linkages between those features.

Very often, the new proposed MPSoCs are driven by a single or very narrow class of applications or just their most dominant part. For example, although

I. Radojevic and Z. Salcic, *Embedded Systems Design Based on Formal Models of Computation*, DOI 10.1007/978-94-007-1594-3_7,

motivation for a new architecture is more general, justification for it is based on relatively simple or restricted case studies such as computation kernels (digital filters such as FIR and IIR, FFT, DCT etc.) or full applications (H.263/4, smart cameras, baseband radio signal processing, OFDM, radar processing, large scale database servers). However, they are typically oriented towards data-dominated systems and almost completely neglect the control-dominated parts of wider applications, multi-modal operations and the need for more complex communication mechanisms when combining those individual parts into larger systems. On the other hand those control-dominated parts, which are usually less computationally demanding, are critical from the application point of view. Examples of such applications with the mixture of control-dominated and data-dominated computations are cellphones and all sorts of portable computers (notably tablets) with the range of applications that can be executed simultaneously or at different times, multi-participant gaming, automotive systems with the range of applications from hard real-time to soft- or non-real-time, home automation, surveillance and tracking, robotics, building automation, industrial automation, smart grid, just to name few.

Also, when addressing the dominant applications of today and of the foreseeable future we have to identify that they typically diverge in two main directions. One is the world of best effort systems, in which the highest performance with possible trade-off with power requirements is the ultimate goal. The second direction is the world of real-time systems, which do not need the highest performance but guaranteed timely response to the requirements of their environment and events generated during system operation, sometimes absolutely uncompromising (hard real-time) or sometimes little bit relaxed when the lack of timely response, from time to time, will not be considered harmful for the system functionality (soft real-time). However, in both types of systems, computations are governed by the interaction with their environment, and the application parts communicate each with the other in ways that can be, or sometimes cannot, specified and analyzed in advance.

The advanced current approaches usually do not consider separation and do not distinguish between control-driven and data-driven components of computation, which typically also have mutually opposing execution requirements: while control usually cannot be reduced or simplified, data-processing can be done with different variations (simpler and alternative algorithms, change of precisions etc.). Complex relationships between control-driven and data-driven parts of applications are dealt with by system designers and programmers in ad-hoc ways with a minimal use of formal modeling and design techniques, if at all, on their individual parts. As the result, complex system design flow is typically broken by a number of manual interventions with significant disconnect between the specification and the real system execution (implementation). The reason for this is lack of formal models used to describe complex systems that combine control-driven and data-driven computations, which was one of the major motivations for the work on DFCharts and then also its extension to DDFCharts. Although not perfect, these MoCs can be used to formally model much bigger class of applications than those used in practice, and at the same time can serve as the central point of system design process and design flow, which can be completely automated, from specification to implementation.

Schetches of this will be given in Chap. 8. While implementation of DFCharts and DDFCharts on single processor computers is rather straightforward by, for example, extending current programming languages with the specialized libraries, we also demonstrate that they can be easily used to describe systems implemented as MPSoCs and specifically target heterogeneous MPSoCs. In this chapter we first describe some of the trends and state of the art in MPSoC development. Practically all existing approaches are oriented towards data-driven (and dominated) applications, although they claim applicability and suitability, and are motivated, by control-driven and multi-mode processing. Then, we introduce a heterogeneous platform that provides explicit processing elements for control-driven processing and interaction with the environment, called heterogeneous reactive architecture (HETRA), suitable to implement wide class of systems which combine control-driven and data-driven processing, including multi-mode systems, in an organized way and satisfy the requirements of complex model of computation such as DFCharts on the execution level. HETRA, as will be seen, adds reactive processors, as new processing elements, to the spectrum of possible processing elements in a heterogeneous execution platform. A typical reactive processor and its integration with other types of processors are illustrated, as this type of architecture is used as a target of mapping DFCharts described systems onto a subset of HETRA execution platform.

7.2 Architecture Framework – HETRA

Majority of MPSoCs are based on using a traditional processor core replicated many times, organized in a network on chip (NoC) and some specific topology suitable for supporting application requirements. The major motivation for NoC approach is scalability, as the networked processors can operate on different and unrelated clocks, which is opposite case to the traditional multiprocessors on chip (CMPs). This naturally goes towards the architectures which are using different, mutually asynchronous clocks and implementation of the architecture as a GALS (Globally Asynchronous Locally Synchronous) system. This fact can be beneficial in addressing models of computation around which the applications are developed. NoCs require specialized routers or switches, which are also replicated to enable connections of, typically, clusters of traditional processors. These clusters can be organized around traditional buses or in some sort of CMP (Chip Multiprocessor). However, NoCs can use circuit switching, packet switching or pipelined links [98] to address varying needs of applications and to address scalability satisfactorily. Still, intelligent use of resources is needed to achieve real scalability, especially at application level [99]. This looks like Lego-like design approach, but the question has to be asked what the bricks are? In order to achieve viable systems a number of accompanying mechanisms are necessary, first of all communication protocols, which have to be identical for the whole system and then result in large overheads. Contrary to usual computer networks, NoC adapters (routers, switches) typically rely on hardware implementation, not software, because of performance requirements and the fact that data exchange happens within chip boundaries. This naturally leads toward communication centric design approach, although very often the system design approach has to be computation centric, too.

The survey [100] nicely summarizes the features of various topologies used for implementation of complex systems on chip, from 2D-mesh technologies, which dominate today's NoCs and are considered most general, to usually more complicated crossbars with non-blocking mechanisms, which may be expensive as the number of communicating components increases; point-to-point connections, single, multi-segment and hierarchical buses, ring crossbars, fat-trees, 2-D torus, and other custom schemes. As the survey shows, some of the approaches are suitable or have been prototyped in FPGAs, and practically all lack of concrete results from real applications, typically only fragments of the applications are prototyped. This clearly indicates that some things are missing, particularly the tools to specify and then map real applications on the NoC as the execution platforms. One of the reasons for this is the lack of tools to describe the applications on a high system level. The designers rely on using existing tools that allow descriptions of only system parts, and then painstaking process of specifying interconnections of these parts on specification level and mapping each individual application, with its own specifics, on the execution architecture manually. This means that the parallelisms, which are very challenging [101], have to be identified and exploited on the application level and then the application mapped on the execution platform. Some partial solutions exist, which allow certain level of customization of the execution platform itself [102], but they still require utilization of standard programming languages for specification. A solution can be to use standard, non-customizable platforms, such as [103], but the question of how to use it with the existing tools remains. The problems emerge from the fact that these platforms offer certain number of processing elements (processors), and the application can use the processor on (1) task (process) level by parallelizing parts, such as typically loops and certain mathematical operations, which again depend on loops, or (2) by using task level parallelism in attempt to distribute execution on multiple cores and hope that it will result in enhanced performance. Scheduling of operations and tasks is critical in this case, and no obvious and general solutions exist. Analysis of designed systems becomes the key, and it is very difficult, if not impossible, if the system is designed in ad-hoc manner without using a systematic design flow based on formal model of computation. Some simple solutions have been proposed, such as allocation of each task to separate (own) processor core [104], which alleviates use of complex scheduling, and transfers the problem to scheduling accesses to memories rather than tasks or operations on individual cores. However, these solutions result in low utilization of processors, especially when their number increases over even moderate threshold (e.g. four processors). Identification of parallelism and its use in embedded and real-time systems context becomes even more difficult, because it may be connected with very complex control and then also constrained with real-time requirements. Furthermore, as analysis often shows, some of the resources in the underlying execution platform are underutilized to a very significant level and their better utilization can be achieved by more customization, e.g. by customizing instruction set of individual general purpose processor cores [103] and specialized processors organized in tightly coupled coarse-grained processor arrays [105], or even more radical by run-time dynamic reconfiguration of processor cores [106] and customization and dynamic reconfiguration of the overall platform [107]. There are many other examples of execution architectures aimed at better fitting to the requirement of fixed or more general applications. Aforementioned solutions are primarily aimed at data-driven applications. There have been also attempts to use reactive processors [108, 109] for control-dominated applications by inventing new multiple core execution platforms [109] for synchronous reactive applications, or even for simple heterogeneous systems [110]. Customization of a processor typically results in low power consumption execution, avoidance of painstaking RTL verification when dealing with hardware implementations, proprietary solutions that protect the IP, automated processes, increased security, avoidance of buses, better area/performance trade-offs, and higher productivity.

It has been demonstrated that homogeneous solutions, which are based on replication of the same computation and communication resources, are easier to implement, but more difficult and less efficient to use for specific applications. The heterogeneity of execution platform can bring many benefits, especially in the domains such as low resource requirements and low power consumption. However, current methods for the exploitation of advantages of heterogeneity are lacking tools and formal mechanisms. The question is how the process from specification to implementation (and execution platform) can become smooth and at each design stage preserve semantics and properties of original specification, including the execution level, while still delivering efficient solutions, optimized for resource and power usage, for example. While formal models of computation can be implemented on any execution platform with the varying degree of design effort, we are seeking for platforms that naturally implement certain model of computation and allow straightforward and automatic mapping from the specification, which complies with the same model, to the implementation. Some of the previously proposed execution platforms have searched for such goals, particularly [110] with the support for synchronous reactive model of computation, and [111] that actually supports asynchronous reactive model. However, both of them were lacking support for data-driven computations and some form of the data-flow model of computation. One concrete example of the architecture that targets GALS model is given in [112, 113], with a limitation that it is customized to the certain degree for GALS model and programming language Systems [114, 115], in which data-driven computations are specified in Java, and as such do not follow any formal model of computation. It demonstrates that the approach of separating control on specialized processors and data-driven processing on traditional processors has its merits and can be further extended to any type of the GALS model regardless of the mechanism used for communication between asynchronous parts of the designed system.

So, what are the options which make execution architecture suitable to support a heterogeneous model of computation such as DFCharts and DDFCharts? First, we need a range of processing elements which cover the needs of applications in terms of both control-driven and data-driven processing. Obvious choices for the later

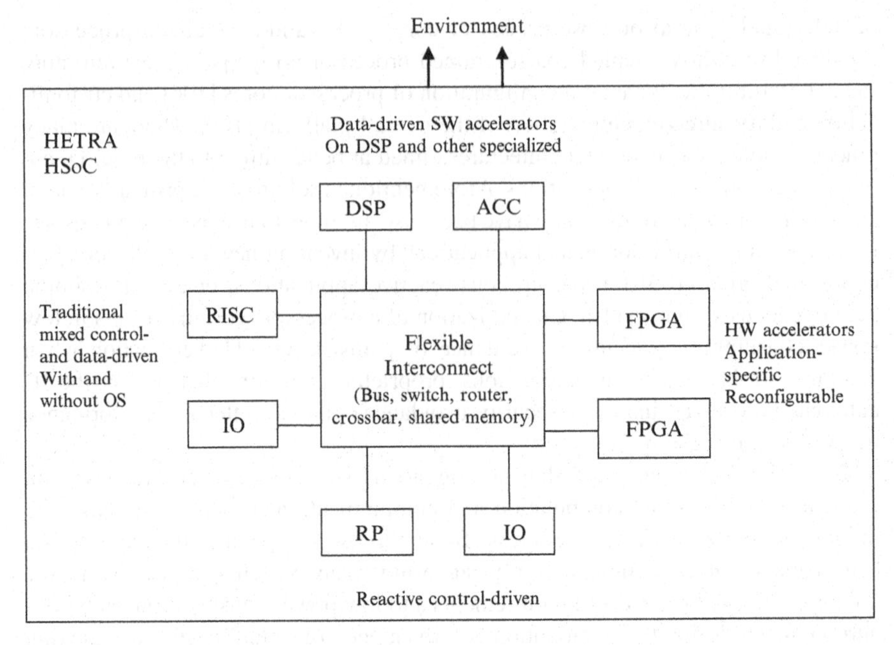

Fig. 7.1 Heterogeneous system on chip (HSoC) – HETRA

are traditional processors (e.g. RISC), their combinations into tightly coupled multiprocessor and multi-core clusters, general-purpose and application-specific digital signal processors (e.g. DSP), and application-specific processors in the form of ASICs and reconfigurable blocks, which can implement data-driven algorithms in hardware (e.g. FPGAs). However, as the result of the need to cover control-dominated part of the applications and efficient implementation of finite state machines and reactivity, we propose inclusion of reactive processors as another processing element for implementation of control part of complex models of computation. Reactive processors also can be efficiently used to implement different scheduling mechanisms and control other processing elements in such systems. As processing elements implement computational functionalities that communicate each with the other, there is a need for a range of communication facilities in the form of flexible interconnect mechanisms, which can be structured and used as the applications need. Then, heterogeneous system on chip (HSoC) is a system on chip that can combine various processing elements and interconnect components to optimally satisfy requirements of an application. A pool of resources from which a customized solution can be implemented is represented in Fig. 7.1. Besides the resources of the heterogeneous platform, the figure shows that the HSoC interacts with the environment and with that indicates the need for reactivity as the feature of the execution platform, which will enable mapping of the reactivity of the heterogeneous model of computation and the specification language. By further narrowing execution platform design options to those needed by the underlying MoC, complex design flow can be significantly simplified. Because the execution platform consists of a number of programmable processors that execute programs, beneficial may be adding of run-time support that helps bridging between model of computation, execution platform and specification (design) language.

The HETRA (Heterogeneous Reactive Architecture) consists of different types of processing elements and flexible interconnect resources which can be combined in custom configurations and concrete architectures to satisfy application requirements. The role of interconnect elements is to provide for efficient message passing and data exchange mechanisms, as fit to the application and processing requirements. Input-output elements have been added, since they are essential in for implementing reactive features, enabling mapping of the real world (environment) onto suitable abstractions of the architecture and the used MoC. Design exploration tools, traditional and more advanced that take into account multi-modal nature of the applications and reconfigurability, are needed to optimize the architecture to a specific one. The use of processing elements and flexible interconnect also takes into account other system requirements and constraints, such as distances between elements, required data bandwidths, reliability and dependability, power and energy, to result in the system that optimally satisfies application requirements.

HETRA enables execution of single and multiple applications at the same time. Applications can consist of multiple logical tasks or behaviors executed on processing elements, which communicate each with the other as the application requires. Those behaviors can be data-dominated or control-dominated, and as such may be more suitable to run on specific processing elements. For example, data-driven tasks are efficiently executed on DSP processors or FPGAs, although they also can be, if less performance-critical, executed on traditional processors (RISC). Control-driven and reactive tasks are primarily run on reactive processors (RP), or if less critical on traditional processors (RISC). The overall reactive application is driven by control-dominated code that naturally executes on reactive processors, while data-driven tasks are primarily executed on specialized processors. The role of reactive processors is critical for the system operation as they are used for implementation of control part of application, which ensures overall system integrity and compliance with the selected MoC. The role of the interconnect mechanism is critical, too, as it provides the necessary low level functionality and performance to the support the data exchange in selected MoC.

7.3 Reactive Processors as the Elements of the Heterogeneous Architecture

Reactive processors as the basic building block of the heterogeneous reactive architectures such as HETRA. Although there exists a number of such processors, e.g. [108, 116–118], we illustrate the basic idea behind them in this section on the example of ReMIC (Reactive MICroprocessor) [108]. We emphasize their capabilities that have been used in mapping DFCharts on specific subset of HETRA architecture, extended version of HiDRA (Hybrid Reactive Architecture) [111]. The ReMIC processor is suitable for handling signals as the physical mechanism to communicate with the external environment and implement reactive behaviors, as well as to deal with preemptions in a structured way. It has special instructions for emission, check of presence and immediate reaction on signal presence and quick transfer of control from one to another behavior (described by FSMs). Also, a variant with power considerations and additional power aware instructions is briefly introduced.

7.3.1 Reactive Microprocessor – ReMIC

ReMIC processor core has been designed to provide:

1. *Efficient mapping of control-dominated applications on processor ISA*: This approach leads to better performance and code density for control-dominated applications.
2. *Support for direct execution of reactive language features*: This ensures that control-dominated models and languages can be supported more directly on a processor without any intermediate code generation performed by conventional compilers.
3. *Support for concurrency and multi-core configurations*: The processor architecture supports building multiple core systems that can simultaneously execute multiple application (concurrent) behaviors and provide a mechanism for their synchronization and communication.

ReMIC design [108] follows the main ideas of reactive model of computation. This is achieved with a set of native instructions in addition to standard instructions found in traditional RISC-type processors. These new native instructions provide direct support for *delay*, *signal emission*, *priority* and *preemption*. Concurrency is achieved by multiple ReMIC cores, where generic signal manipulation instructions can be used to implement efficiently synchronization of concurrent tasks running on separate processors. The key features of ReMIC that facilitate reactive applications are summarized as follows:

- One Signal Input Port (SIP) and one Signal Output Port (SOP) are implemented to enable direct mapping of pure (binary), effectively enabling processor to communicate with its environment by direct manipulation of values represented on wires. Simultaneous emission of multiple signals in one clock cycle is also supported.
- A number of user programmable internal timers are implemented to generate the timeout signals, which can be fed back and used internally for synchronization purpose or used to interact with the environment.
- ABORT instruction is introduced to handle preemption in a structured way, typical for reactive programming languages. Code can be wrapped up in the abort

Table 7.1 Reactive instruction set

Feature	Instructions	Semantics and descriptions
Signal manipulation		
Signal emission	EMIT *signal(s)*	*Signal(s)* is/are set high for one tick.
Signal sustenance	SUSTAIN *signal(s)*	*Signal(s)* is/are set high forever.
Delay	TAWAT *delay*	Wait until *delay* (number of instruction cycles) elapses.
Signal polling	SAWAIT *signal*	Wait until *signal* occurs in the environment.
Conditional signal polling	CAWAIT *signal1, signal2, address*	Wait until either *signal1* or *signal2* occurs. If *signal1* occurs, execute instruction at the address immediately followed, or else at the specified *address*.
Signal presence	PRESENT *signal, address*	Instruction at the address immediately followed will be executed if *signal* is present, or else at the specified *address*.
Preemption		
Preemption	ABORT *signal, address*	Program finishes its current instruction and jumps to *address* in the occurrence of *signal*.

statement and immediately abandoned when an external event on the specified SIP input occurs. A customizable number of levels of nested aborts is supported with maximum one instruction cycle reaction time on external signal. Besides the execution efficiency and determinism, this also leads to very structured programming style when programming the reaction on external events.
- Other instructions including EMIT, SUSTAIN, PRESENT, SAWAIT, TAWAIT, CAWAIT are added to support reactive behaviors (refer to Table 7.1 for a list and meaning of these instructions).

Figure 7.2 illustrates the ReMIC partition of functionality and mechanism for connection with external world. It consists of a traditional RISC-type pipelined microprocessor datapath, reactive functional unit (RFU) for handling external and internal signals, and processor control unit. ReMIC has Harvard architecture with 32-bit wide program and 16-bit wide data memory, which both are also parameterized in the design and can be changed. Although we will not concentrate on its customization features here, ReMIC is a parameterized soft-core that can be instantiated in an application-required configuration and also easily extended with the additional instructions and corresponding functional units for their execution.

Generally, the hardware implementations of reactive instructions result in better performance and much more efficient compilation of reactive programs [118]. All ReMIC instructions have same length, 32 bits in default configuration. Instruction formats of some reactive instructions are illustrated below.

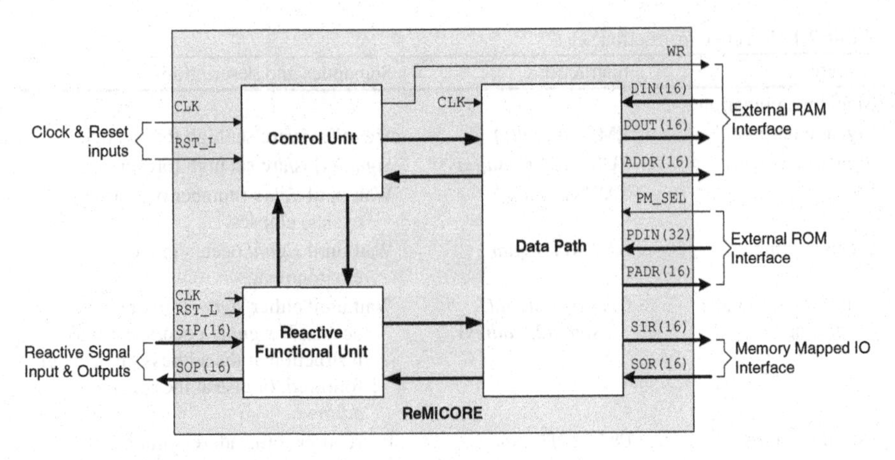

Fig. 7.2 REMIC block diagram

7.3.1.1 EMIT – Signal Emission

EMIT, with the format as shown below, is used to generate external output signals through the signal output port (SOP). The signals last for one clock cycle. Bits 24–9 of the instruction are mapped to bits 15–0 of the SOP.

31–30	29–23	24–9	8–0
AM(2)	OC(5)	Signals(16)	Unused(4)

7.3.1.2 SAWAIT – Signal Polling

SAWAIT, with the format as shown below, is used to poll for a specified signal from the signal input port (SIP). ReMIC stays in a wait state until the signal occurs in the environment.

31–30	29–23	24–9	8–5	4–0
AM(2)	OC(5)	Unused(16)	SIG(4)	Unused(4)

7.3.1.3 ABORT – Preemption

ABORT, with the format as shown below, is the most crucial reactive instruction because it is introduced to support preemption with priorities. An ABORT instruction has a signal, which is sensitive to it and a continuation address. ABORT instruction becomes active from the instant it is executed until either (1) it reaches the continuation address, or (2) an event on one of the SIP inputs occurs that preempts all unexecuted instructions within the body. Bits 24–9 of the instruction specify the abort continuation address and bits 8–5 specify the abort signal that is encoded to one of the SIP inputs.

31–30	29–23	24–9	8–5	4–0
AM(2)	OC(5)	Continuation Address(16)	SIG(4)	Unused(5)

The idea behind practically all reactive processors [108, 116–119] is inspired by Esterel [15], a synchronous reactive language specifically designed for reactive programs. Esterel provides a set of constructs for modeling, verification and synthesis of reactive systems.

7.3.2 Power Aware ReMIC-PA

While most research related to reactive processors focus on improving the performance, not enough attention have been made in analyzing power consumption, which is obviously very important in many embedded applications and then also on MPSoCs and HSoCs in which they will be used. ReMIC has been analyzed and then empowered with mechanisms that reduce its power consumption and enable power aware applications. Power aware ReMIC, or ReMIC-PA [120], is implemented through power-aware optimizations applied to ReMIC. The modifications target reducing dynamic power dissipation by minimizing switching activity of the design. The optimizations included modifications that can be classified into two parts, targeting data-dominated and control-dominated applications, respectively.

There are two optimization techniques which can be effective for data-dominated parts of applications: (1) Precise Read Control (PRC) and (2) LiVeness Gating (LVG) [121]. The PRC is used to eliminate the register file reads based on instruction types. In the original ReMIC design, every instruction automatically reads two operands from register file in the decode stage no matter what instruction it is. This mechanism facilitates the proper pipeline operation at the cost of unnecessary reads. For example, instructions that operate on immediate values do not need to read the second operand from the register file. The LVG is responsible for elimination of the register file reads and pipeline registers updates when the pipeline is stalled or a taken branch instruction is executed. It is obvious that in above two situations, both register file reads and pipeline registers update are worthless and should be eliminated. To support PRC and LVG, the extra circuitry used to control the register file access and pipeline register update is inserted into the original ReMIC control unit. The operation codes of instructions are also reordered according to the instruction types.

The optimizations for control-dominated applications are focused on minimizing the switching activity of the clock signal. ReMIC-PA provides a mechanism, which suspends the system clock when the core is idle and restores it when the designated input signals fed by the environment occur, to reduce power dissipated by the clock

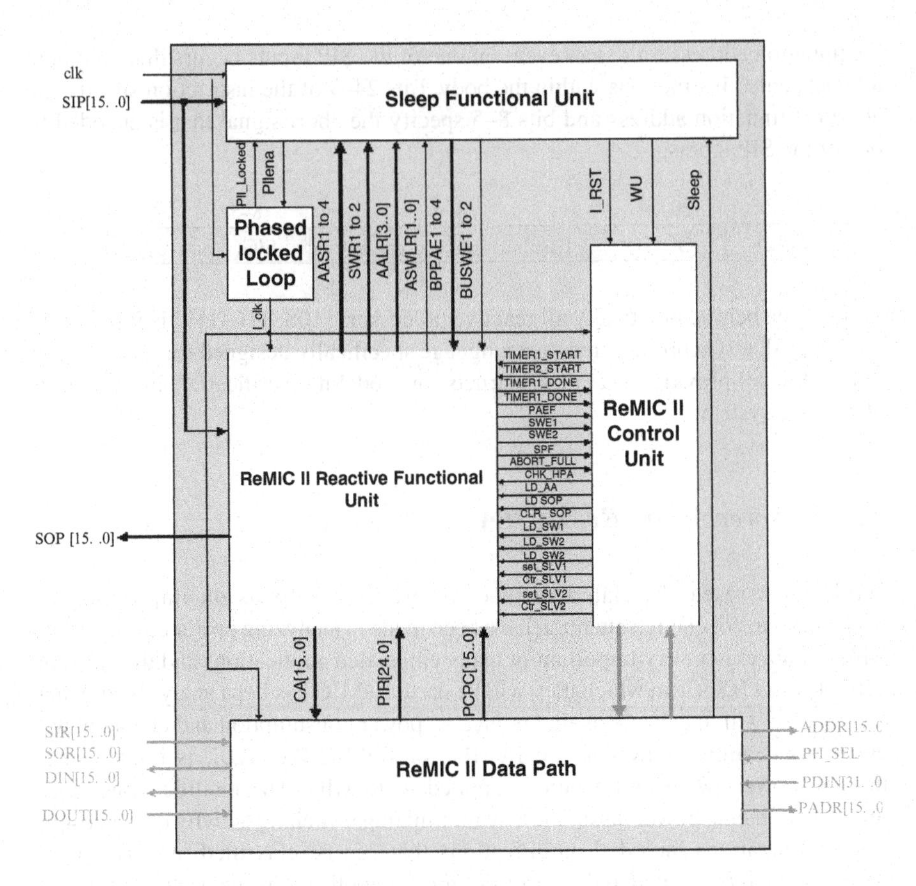

Fig. 7.3 ReMIC-PA block diagram

signal. To support this mechanism, a phased locked loop (PLL) and a functional unit, called power control functional unit (PCFU) have been added to the original ReMIC. Figure 7.3 illustrates the modified architecture of ReMIC-PA.

ReMIC -PA provides three architectural supports for power optimizations:

1. Two execution modes, called the *normal* mode and the *sleep* mode. In the normal mode, the PCFU enables the PLL to produce the system clock so that ReMIC-PA operates similar to ReMIC. In the sleep mode, on the other hand, the PLL is turned off by the PCFU so that the system clock is gated and ReMIC-PA is suspended. The transition from the normal mode to the sleep mode is carried out by the execution of sleep mode related instructions. The restoration from the sleep mode to the normal mode is activated by the designated external input signals.
2. The PCFU runs at the lower frequency than the processor core does. As shown in Fig. 7.3, the clock fed to the PCFU is the same as the reference input clock of the PLL. Although the input clock frequency of the PLL can be an arbitrary value allowed by the FPGA device used in prototyping, in this case it is lower than the output clock frequency to reduce power dissipated by the PCFU itself.

Table 7.2 ReMIC-PA additional power control instruction set

Features	Instruction syntax	Corresponding reactive instruction	Function/description
Power-efficient signal sustenance	LSUSTAIN *signal(s)*	SUSTAIN	Bring the processor to the sleep mode and set signal(s) high forever.
Power-efficient signal polling	LSAWAIT *signal*	SAWAIT	Bring the processor to the sleep mode and wait unit the specified signal occurs in the environment.
Power-efficient conditional polling	LCAWAIT *signal1, signal2, address*	CAWAIT	Bring the processor to the sleep mode and wait until either signal1 or signal2 occurs. If signal1 occurs, the processor is restored to the normal mode and executes instruction from consecutive address; or else from the specified address.
Suspend	AWAIT	NONE	Bring the processor to the sleep mode.

3. A set of power-efficient reactive instructions is provided by ReMIC-PA to support power management and optimizations explicitly by the system designer.

ReMIC-PA has four power-efficient instructions that facilitate the mode transition. All instructions are presented in Table 7.2. They are 32-bit long and follow the standard ReMIC instruction format.

Operations performed by the power-efficient instructions are almost the same as that of the corresponding reactive instructions, except that the reactive instructions bring the processor in a wait state while the power-efficient instructions cause the processor to be suspended. These new instructions can be mixed with the other reactive instructions. When external input signals with shorter response deadlines are considered, the reactive instructions are used; otherwise, the power-efficient instructions are preferred.

7.4 Example of Heterogeneous Reactive Architecture – HiDRA

In this section we introduce an architecture that has been used to implement HSoCs, in this particular case application with DFCharts as an underlying MoC. HiDRA (Hybrid Reactive Architecture) [111] is proposed and made for rapid prototyping and implementation of heterogeneous embedded systems based on a set of reactive processor cores, ReMICs, but also allows mixing with traditional microprocessors and application specific hardware implemented algorithms and behaviors. As such HiDRA can be considered as an example of HETRA approach.

7.4.1 An Overview of HiDRA

HiDRA is a heterogeneous system that consists of multiple reactive processor cores for control-dominated software-implemented behaviors and functional units and traditional processors data-driven hardware-implemented and software implemented behaviors, respectively. It allows interconnecting hardware- and software-implemented behaviors in almost arbitrary way, where they synchronize and communicate each with the other using signals (essentially wires) and for exchange of data use distributed, shared and multi-port memory blocks. The architecture is suitable for FPGA prototyping as it uses some of the features of current FPGA devices like distributed SRAM memory blocks, but easily fits to the HSoC approach and ASICs. The architecture has the following major features:

1. It allows execution of concurrent behaviors that can be software-implemented and run on a number of physical ReMIC processor cores, traditional processors and/or hardware-implemented in the functional units.
2. It can be run in two types of implementations, one with a master processor and another without any master. Here we will describe the case where one reactive processor assumes the role of a master processor. A master processor (MP) performs system initialization and coordination of concurrent activities, but also can implement application behaviors.
3. Other processor cores, called application processors (AP), are used to implement only concurrent application behaviors.
4. Behaviors implemented on reactive processor cores directly support reactive, FSM-type model of computation, in addition to declarative-type computation used in data-driven applications in traditional processors.
5. Concurrent processes implemented in hardware functional units and traditional processors are primarily aimed at data-driven behaviors.
6. Communication between behaviors, whether they are implemented using programs that run on processor cores or hardware-implemented functional units, is achieved by means of a layer of shared memories (sometimes just registers), which connect processor cores and functional units as illustrated in Fig. 7.4 and can be considered an implementation of flexible interconnect of HETRA approach shown in Fig. 7.1.

It should be noted and will be obvious from the following presentation that in principle there need not to be any processor declared as the master processor and it is not requirement of HiDRA approach. However, in the examples of use of HiDRA in DFCharts implementation shown in Chap. 8 we assume this type of HiDRA configuration.

Figure 7.4 illustrates the major HiDRA concepts and shows an example of possible mappings of an application represented by the dependence graph of behaviors

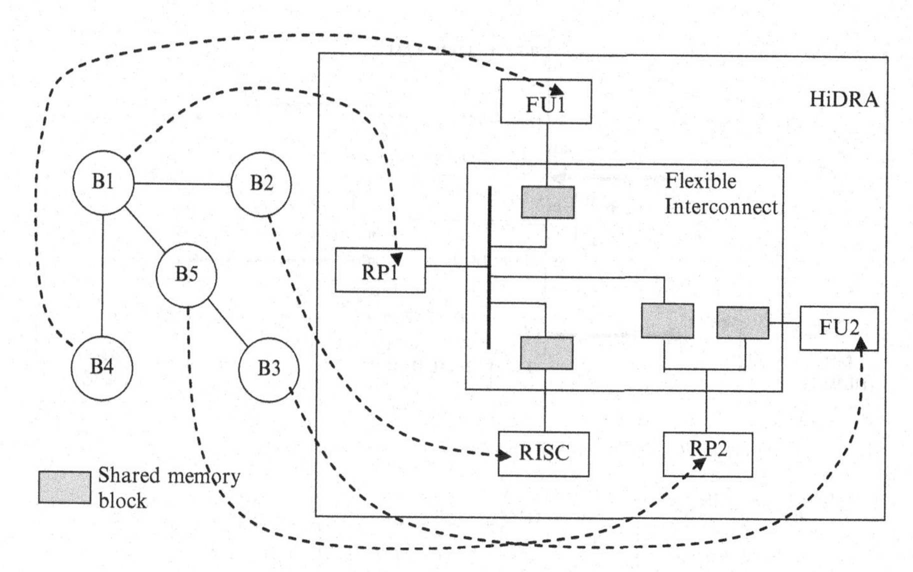

Fig. 7.4 Mapping of an application to HiDRA architecture

(B_1,…,B_5), which can be also considered as application tasks or processes, on the architecture. HiDRA in Fig. 7.4 has a hierarchical topology that contains one master processor (MP) which implements main system behavior in software and has capability to control all other units (processors and functional units, FU, called with a single name application processors) that implement other behaviors. Application processor (AP) behaviors can activate hardware-implemented behaviors on their local functional units (e.g. RP2 and FU2). Synchronization between behaviors is accomplished via signal manipulation statements that are explained in the further text (which emit a signal or poll a signal for the presence of an event). Communication primitives are implemented using shared memories and signal manipulation statements that perform low-level handshaking. A range of communication primitives such as bounded FIFO and rendezvous are also supported.

7.4.2 An Implementation of HiDRA

Here we present an implementation of HiDRA in ALtera FPGA device. While ReMIC microprocessors are implemented using logic elements, shared memory (SM) is implemented using a triple-port SRAM module from the Altera megafunction library, which has two read ports (Qa, Qb) and one write port (Din). The MP is configured with read-and-write access to one SM and read-only access to the other, as shown in Fig. 7.5, while the AP is configured the other way around. The data memory maps of the MP and APs are configured as shown in Fig. 7.6. The control signals between the MP and APs are configured as shown in Fig. 7.7. The MP uses four SOP signals and four SIP signals for each AP and spares the same number for connection with external environment.

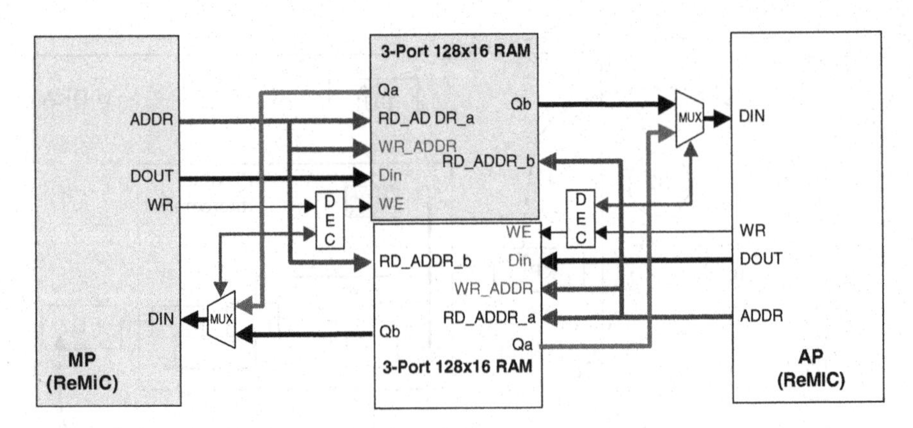

Fig. 7.5 SM configuration between MP and AP

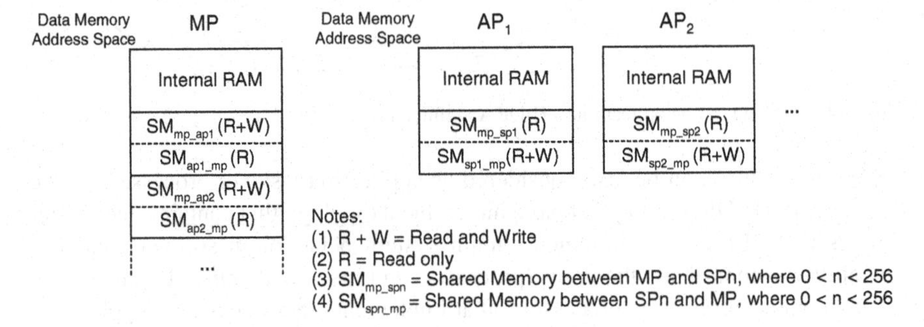

Fig. 7.6 Data memory address spaces of MP and SPs

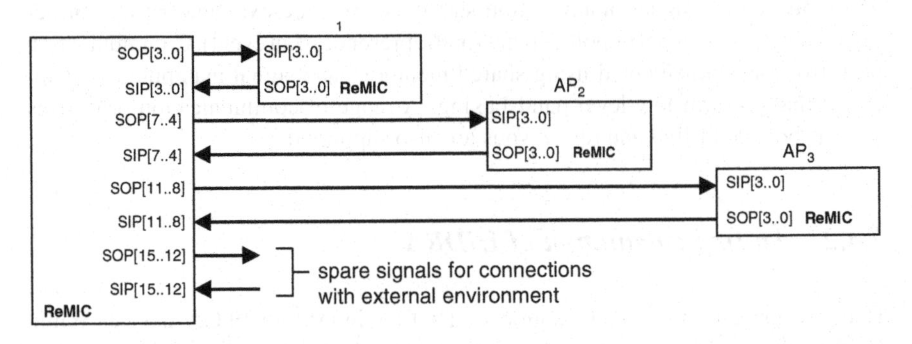

Fig. 7.7 Signal connections between MP and APs

Figure 7.8 illustrates a complete HiDRA implementation with one MP and two AP reactive processors using the above configurations. Shaded area in the middle actually represents the implemented flexible interconnect.

HiDRA gives a number of mechanisms that can be used to implement blocking and non-blocking read and write when necessary. As signal lines are used for

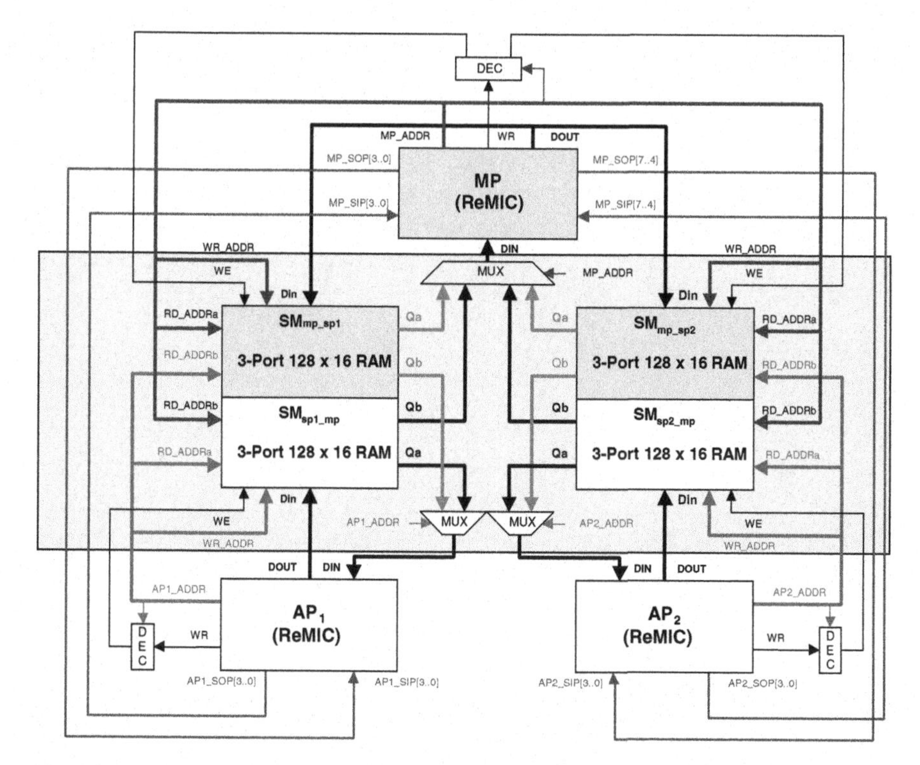

Fig. 7.8 HiDRA implementation with one MP and two APs

synchronization between processors, they can be manipulated directly from the corresponding ReMIC reactive instructions, which all perform in a single instruction cycle. To emit a signal and notify another processor on the event requires single instruction cycle, while the notification is received in either a single cycle (if using ABORT as non-blocking mechanism) or in a number of cycles depending on incoming event (when using AWAIT instructions as blocking mechanism).

Similar approach is used when connecting standard microprocessors to HiDRA. Shared memories are typical interconnect mechanism for data exchange. Instead of shared memories sometimes registers are sufficient to implement required communication bandwidth.

Chapter 8
Implementation of DFCharts on HiDRA

Previous chapters deal with system level by exploring issues such as specification, simulation and formal semantics of DFCharts without any reference to implementation. The focus of this chapter is implementation of DFCharts specifications. The target architecture for implementation is HiDRA, which was described in Chap. 7. HiDRA is capable of implementing both control-dominated and data-dominated types of behaviour that are found in DFCharts. It has special features supporting reactivity while data-dominated operations are supported using traditional solutions. In principle, any multiprocessor architecture may be used for implementation of DFCharts. However, because of special features that support reactivity, HiDRA is likely to provide more efficient implementations. This is the main reason for selecting HiDRA. In Sect. 8.1, we present a design methodology for implementing DFCharts on HiDRA. It consists of five steps: specification, FSM compositions, allocation and partitioning, synthesis and performance evaluation. We are mainly concerned with the synthesis step, which is separately treated in Sect. 8.2 where details regarding the execution of DFCharts on HiDRA are presented. In Sect. 8.3 we show how the methodology is applied on the frequency relay case study.

8.1 DFCharts Design Methodology

The design methodology is presented by the design flow that consists of five steps shown in Fig. 8.1. The first step is to make the DFCharts-based specification of a system without any reference to HW/SW implementation. Before the specification is mapped onto the implementation architecture, the designer has an opportunity to merge multiple FSMs into a single equivalent FSM by parallel and hierarchical compositions.

I. Radojevic and Z. Salcic, *Embedded Systems Design Based on Formal Models of Computation*, DOI 10.1007/978-94-007-1594-3_8,

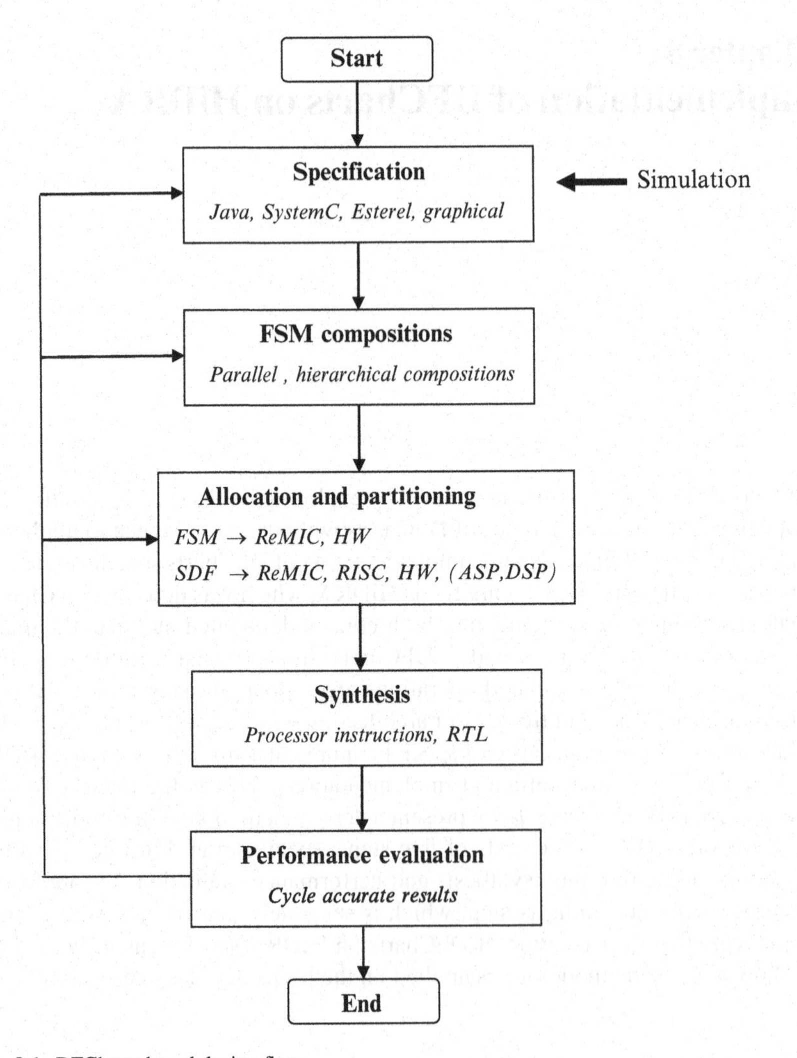

Fig. 8.1 DFCharts based design flow

In the third step, an instance of HiDRA is created by connecting ReMICs and functional units. Then, FSMs and SDFGs are partitioned among the processing elements. As this step is difficult to automate, the designer can freely explore different implementation options. A complete implementation is created by automatically synthesizing FSMs and SDFGs into ReMIC instructions and RTL in the fourth step. Its performance is evaluated in the fifth step. If the implementation is not satisfactory, the designer can return to any of the first three steps.

A key feature of the methodology is that functionality and implementation are completely separated. Functionality is only verified at the specification level, using simulation and formal verification.

8.1.1 Specification

Specifications are created in Java environment described in Chap. 6. Currently, textual design entry is only supported, but a graphical user interface can be added later, which would enable specifications to appear as in Chap. 3. It is also possible to employ SystemC or Esterel for creating specifications after modifying them according to the guidelines given in Chap. 5. It is important to note that when an SDFG is created, its relative speed is specified by indicating how many ticks of the FSM clock each iteration takes, as pointed out in Sect. 6.1.2.

As indicated in Fig. 8.1, validation can be done using simulation. Simulation of DFCharts-based designs in Java has been described in detail in Chap. 6. While simulation is the most widely used means of validation, formal verification has been gaining importance. Formal verification of DFCharts has not been discussed in this book. An efficient method for formal verification of DFCharts can be developed using the MCFSM model. This will be an important future research direction.

8.1.2 FSM Compositions

There is overhead associated with running an FSM on a processor or functional unit. For example, variables must be used to remember the current state of an FSM and indicate if it has made a transition in the current tick. Local signals that are used for communication between FSMs always consume resources even if the FSMs are mapped onto a single processing element. When an FSM composition is applied on two FSMs, they will be executed as a single equivalent FSM. Two important effects are produced as a result. The overhead is reduced, since there is now one FSM instead of two, and local signals are removed. Thus, FSM compositions give the designer an opportunity to trade concurrency for more efficient execution. Two types of FSM compositions are used: parallel and hierarchical. The two compositions and their relation to the synchronous parallel and hierarchical operators will be discussed in more detail in Sect. 8.2.3. At this point it is important to stress that the parallel composition produces an exponential increase in the number of states, while the hierarchical composition produces a linear increase in the number of states. For example, looking at the frequency relay specification, if FSM5 (four states) and FSM6 (two states) from Fig. 3.8 are merged by the parallel composition, the resulting FSM has eight states. If FSM2 (three states) from Fig. 3.9 is merged with FSM7 (seven states) from Fig. 3.10 by the hierarchical composition, the number of states in the resulting FSM increases linearly to 9.

8.1.3 Allocation and Partitioning

An instance of HiDRA is created by connecting ReMICs and hardware-implemented functional units. Communication between processing elements is performed through shared memories and ReMIC signal lines. How many processing elements are allocated depends on the desired performance. A higher number of elements may provide better performance, but the cost increases at the same time.

An SDFG can be mapped on a ReMIC or functional unit. Traditional processors, digital signal processors (DSP) or application specific processors could also be included in the architecture for SDFG implementation as long as they satisfy some simple interface requirements. Multiprocessor implementation of a single SDFG is possible [23]. Even mixed HW/SW implementation for SDF has been demonstrated [53]. We could easily incorporate those techniques in our methodology. However, we will assume that each SDFG is executed by a single processing element in order to simplify the discussion. FSM can also be mapped on a ReMIC or functional unit implemented as a pure digital hardware. Although, performance boost provided by hardware is more likely to be needed for computationally intensive SDFGs, FSMs and SDFGs cannot be mapped on the same element. The reason for this restriction is due to different implementation requirements posed by FSMs and SDFGs. FSMs react to events and perform minor computations. Reactivity is not important for SDFGs, but computations are intensive. By executing only one type of behaviour, processing elements can be customized to do their tasks efficiently with minimum resources.

ReMIC consists of three parts: reactive functional unit (RFU), control unit and datapath. If it executes an SDFG, RFU becomes unnecessary. When RFU is removed, ReMIC becomes a Harvard-type microprocessor with RISC-type pipelined datapath, suitable for transformational operations performed by SDFGs. On the other hand, an FSM needs RFU but the datapath does not need to be as powerful. ReMIC has a customizable register file with up to 16 registers. If it implements an SDFG it may need all 16 registers. In the case of FSM implementation, just two might be enough. Furthermore, pipelining, which is useful for predictable SDF data transformations, could be redundant for FSMs.

When FSMs are partitioned among multiple processors, the key consideration is how to maximize the utilization of the processors. We say that a *global tick* is completed when each FSM in the system makes a transition (completes its *local tick*) by computing outputs and setting the next state. Because of the lock-step execution assumed at the specification level, it must not happen that an FSM executes two transitions in a sequence while another one executes none. When all FSMs on a single processor have finished their local ticks, the *processor tick* has been completed. At this point the processor suspends execution and waits for the remaining processors to complete their ticks, and a new global tick can begin. In the ideal case, all processor ticks take equal time which means that all processors are fully utilized, which is rather difficult to achieve.

Two factors are important for maximizing processor utilization: distribution or balancing of loads and inter-processor communication. By loads, we mean execution times of FSM transitions. Loads should be distributed as evenly as possible. In any state of the system, the sum of execution times of FSM transitions should not differ largely across the processors. This may be a difficult task considering that the system may have a large state space resulting in many combinations that have to be explored. However, the designer can be assisted by profiling tools that indicate the amount of time consumed by various execution paths. At this stage, before the synthesis has been completed, only approximate times can be provided. Cycle-accurate performance results are examined in the fifth step.

Balancing of loads is additionally complicated by communication dependencies between FSMs that are executed by different processors. It may happen that a transition of an FSM cannot be completed because it depends on a transition of another FSM that is executed on another processor. Therefore, it is desirable to locate FSMs that communicate by signals on a single processor.

Among the processors executing FSMs, one must be designated as *master processor*. The others are called *slave processors*. Apart from running its FSMs, the master processor has the additional tasks of transferring data across rendezvous channels, controlling the execution of SDFGs, and managing global ticks.

Mapping SDFGs takes less effort. SDFG inputs usually arrive in regular intervals from the external environment. The main issue is whether a processor executing an SDFG is fast enough to complete an iteration before next inputs arrive. Also, it is possible to map multiple SDFGs on a single processor, if it is fast enough to service all of them.

8.1.4 Synthesis

A complete implementation is obtained by synthesizing ReMIC instructions, RTL for functional units, and communication between processors which is realized by ReMIC signal lines and shared memories. The program memory of a ReMIC executing FSMs consists of three sections: FSM threads (FT), FSM scheduler, and tick handler. If SDFGs are executed, the program memory consists of two sections: SDF threads (ST) and SDF scheduler. An FSM thread implements the functionality specified by an FSM. An SDF thread implements the functionality specified by an SDFG. It simply consists of the code that implements SDF actors. SDF scheduler, FSM scheduler and tick handler can be considered as 'middleware' that enables threads to run. An SDF scheduler invokes SDF actors according to a static schedule. It also has to implement a simple interface with the master processor. In Sect. 8.2 we will not deal with the SDF scheduler, since it is described in detail in [23] and many other publications that followed. An FSM scheduler runs FTs in a round robin fashion according to a statically determined schedule. It is possible that an FSM thread is not ready to make a transition when picked by the scheduler due to unresolved local signals. When the scheduler encounters such FSM thread, it selects the next FT, but it will come back in the next cycle. A signal is unresolved if the thread that writes it has not decided yet whether it is present or absent in the current tick. Obviously, a thread attempting to read an unresolved signal cannot proceed. The scheduler has to repeat the schedule until all FSM threads have made a transition. The control then goes to the tick handler. Section 8.2 shows the general flow of control between FTs, FSM scheduler and tick handler. In hardware-implemented functional units, concurrent executions are possible without the need for schedulers (Fig. 8.2).

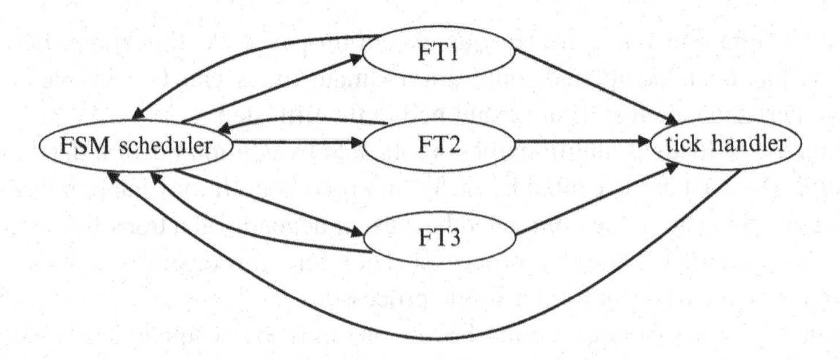

Fig. 8.2 Flow of control between FTs, FSM scheduler and tick handler

The scheduler and FSM threads appear exactly the same on both the master and slave processors. However, tick handlers are different. When a slave tick handler gets control, it informs the master processor that all of its FSM threads have completed a transition. When the master tick handler receives this message from all slave processors the global tick is completed. The master tick handler instructs all slave tick handlers to update shared variables and clear written signals before the next global tick begins.

8.1.5 Performance Evaluation

Although estimations in step 3 could provide useful feedback about a particular implementation, cycle accurate results obtained after synthesis are needed to show precisely whether various constraints are satisfied. A typical requirement in a DFCharts-based specification is to ensure that an SDFG does not miss any input samples from the external environment. According to the DFCharts semantics, the communication on external SDF channels is performed by rendezvous as on the internal channels. However, due to the nature of embedded systems, the external environment will not really wait. If an SDFG is not ready to receive a sample when it arrives, the sample will be inevitably lost unless it is buffered. For many applications, though, buffering on input channels is not a satisfactory solution since it may result in unacceptable delays.

8.2 Execution of DFCharts Specifications on HiDRA

8.2.1 Signals and Variables

Signals and variables enable synchronization and communication among code sections. The most important difference between signals and variables is in their physical implementation. Variables are always implemented with memory.

Signals can be implemented with memory or ReMIC signal lines that are linked with special instructions supporting reactivity.

Signals can either be visible in the specification or inserted during synthesis. The former will be called specification-visible signals, while the latter will be called specification-invisible signals. The specification-visible signals can be input, output or local. Furthermore, we distinguish between external and internal signals. An external signal is used for communication between a processing element and the external environment. An internal signal is used for communication between two processing elements. External signals include specification-visible input and output signals. Internal signals include specification-visible local signals and specification-invisible signals. Specification-invisible signals are listed below. Some signals that are involved in the same type of operation are grouped together and only briefly described. Their use will become obvious in later sections.

1. FT_start: Written by an FT upon entering a hierarchical state. Read by FTs that refine the hierarchical state. May be written and read any time during the tick execution.
2. FT_stop: Written by an FT upon leaving a hierarchical state. Read by FTs that refine the hierarchical state. May be written and read any time during the tick execution.
3. ST_start_request: Written by an FT upon entering a hierarchical state any time during the tick execution. Read by the master processor's tick handler at the end the global tick.
4. ST_stop_request: Written by an FT upon exiting a hierarchical state any time during the tick execution. Read by the master processor's tick handler at the end of the global tick.
5. ST_start: Written by the master processor's tick handler at the end of the global tick. Read by an ST. Unlike the signals described above, this type of signal is held high briefly, just long enough for the ST to respond.
6. ST_stop: Written by the master processor's tick handler at the end of the global tick. Read by an ST. Like the ST_start signals, this type of signal is a short pulse.
7. FT_rendezvous_ready: Written by an FT upon entering a rendezvous state any time during the tick execution. Read by the master processor's tick handler at the end of the global tick.
8. ST_rendezvous_ready: Written by an ST at the end of an iteration. Read by the master processor's tick handler at the end of the global tick.
9. Rendezvous_done: Written by the master processor's tick handler at the end of the global tick when a data transfer between shared memories has been done. Read by an FT any time during the tick execution and by an ST between two iterations.
10. Tick_finished, Start_tick_end_actions, Tick_end_actions_completed, Start_tick: Read and written by tick handlers at the end of the global tick. The purpose of these signals is to synchronize tick handlers so that certain actions are done in the right order before the next global tick begins.
11. Update_shared_variable: Written by the thread that writes the shared variable. Read by the master processor's tick handler at the end of the global tick.

a

```
LDR R0 m
LDR R1 #b
CLFZ            -- clear zero flag
AND R2 R0 R1
SZ              -- skip if zero
```

testing a memory bit

b

```
PRESENT r
```

testing a reactive signal

Fig. 8.3 Comparing memory and reactive signals

Two bits are needed to implement every internal signal that can be written and read any time during the tick execution. Signals that need two bits are specification-visible local signals. Among specification-invisible signals, FT_start and FT_stop signals need two bits. In each case, one bit is needed to indicate whether the signal has been resolved while the other one shows the status of the signal. If writing and reading threads are on the same processor the bits are contained inside the local memory. If they are distributed on multiple processors then the bits must be implemented by input/output registers. Both solutions may be needed depending on how threads are allocated. Currently memory mapped registers *sir* and *sor* are used. With minor modifications of ReMIC, the registers *sip* and *sop* that implement reactive signals can be employed. The modifications would ensure that an emitted signal is held high for the duration of the whole tick instead of just one processor cycle. In the current form, reactive signals can be used just for external input and output signals.

If ReMIC's reactive signals were used for implementation of internal signals, the code size would be reduced significantly since fewer statements are needed for reading and writing by using *emit* and *present* instead of ordinary instructions. The difference between using reactive signals and memory is shown in Fig. 8.3. Part (a) shows that five instructions are needed for testing a bit in memory *m*. The test is performed by ANDing the memory contents loaded in register R0 with the bit pattern *b* loaded in register R1. If the result is zero then the bit under test is zero. Part (b) shows that only one instruction is needed for testing reactive signal *r*.

We can divide variables into specification-visible and specification-invisible, as we did for signals. A specification-visible variable can be a shared variable that is used by multiple FSMs, or local variable that is used by a single FSM. The main specification invisible variables are listed below:

1. *next_thread*: It points to a code section in the FSM scheduler. It indicates which FT the FSM scheduler will attempt to run next.
2. *next_state*: Each FT has this variable. It indicates which state the FT will take in the next tick.
3. *pc*: (program counter). Each FT has this variable. It points to a section of code in an FT.
4. *threads_done*: Each bit in this variable is used to indicate whether an FT has completed its local tick in the current global tick.

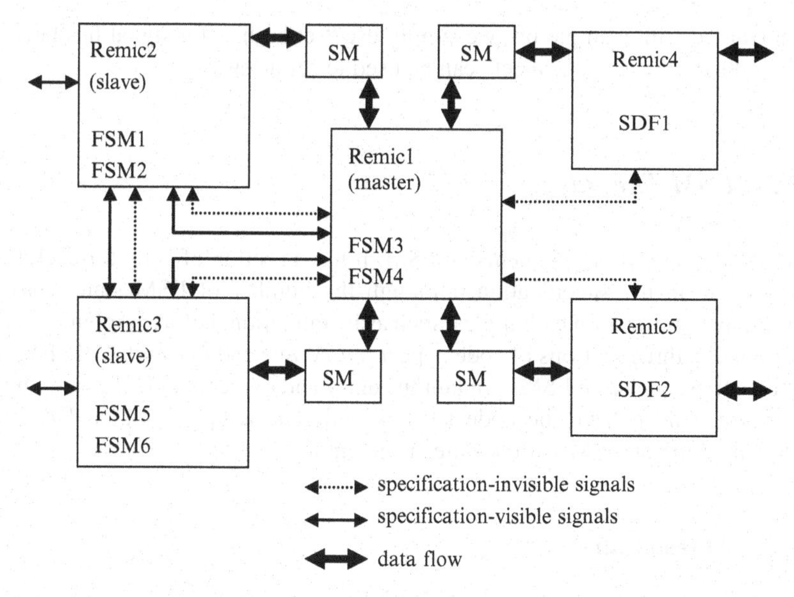

Fig. 8.4 Architecture for DFCharts implementation

For each specification-visible shared variable, two locations in memory are needed. One location is needed for the current value of the shared variable while the other is needed for storing the value that the shared variable will take in the next tick. If the writing thread and the reading threads are all executed on a single processor, the locations needed for the shared variable will be in the processor's local data memory. Otherwise the shared memory which is accessible by the master processor must be used. When transferring data from one shared memory to another, the master processor reads a value from one shared memory into a register, and then writes the value to the other shared memory using the same register.

The flow of data together with signals connections for a configuration of five ReMICs that implement six FSMs and two SDFGs is shown in Fig. 8.4. When large amounts of data are involved in transfers between shared memories, a DMA controller may be a useful addition to the architecture, which would enable data to flow directly between shared memories instead of going through the master processor.

Before describing FSM thread, we explain the notation and terminology that we will use. In figures that present implementation templates, variable names are in italics. Signal names are not italicized, but they always begin with an upper case letter. Constant always begins with an underscore as in _number_of_slave_processors. We use brackets when we want to relate a signal to a particular object like FT, state, channel etc. For example, FT_stop(ft1) means that an FT_stop signal is used by the FSM thread ft1. Labels for code sections are underlined. For internal signals that are implemented by two bits, we use "emit" and "resolve" to denote setting high the status and resolution bits respectively. For internal signals implemented with a single bit, we simply use "set high". If a signal has been set high in the current tick, "cancel" can be used to set it low.

8.2.2 *FSM Thread*

An FSM thread (FT) implements an FSM. It has a section of code for each FSM state visible in the specification, although the number of FSM states may be reduced by optimization techniques, such as bisimulation, before implementation. Additionally, three sections of code appear: *entry*, *exit* and *tick end*. Only FTs that implement preemptable FSMs contain the entry and exit code. On the other hand, every thread must have the code for tick end. The template for an FT called *ft* implementing an FSM with *m* states is given in Fig. 8.5.

8.2.2.1 Thread Entry

A preemptable FSM thread ft is initially in ft entry. The signal FT_start(ft) determines whether ft will start by entering the initial state s_1. FT_start(ft) is written by the FT which has a state refined by *ft*. If FT_start(ft) is not resolved the control will jump to a location in the scheduler which is pointed by the *next_thread* variable. *pc* will still be set to ft entry. However, ft has not finished its local tick which means that the scheduler will have to return to ft before the next global tick can begin. If FT_start(ft) is present, the next state will be the initial state. If the initial state is hierarchical, FT_start signals are emitted and resolved for all FTs that refine the initial state, and ST_start_request signals are set high for all STs that refine the initial state. If FT_start(ft) is absent, ft will remain in ft entry. When FT_start(ft) is resolved the local tick is completed after executing the necessary instructions, so the control jumps to ft tick end. It does not go to ft exit since an FT cannot be entered and exited in the same tick. The other way around is possible though. An FT can be exited and entered in the same tick when the hierarchical state that is refined by it makes a transition back to itself.

It is important to notice here that ft has to complete its local tick even though it has not yet entered the initial state that is visible in the specification. From the point of view of the tick handler, each thread is always active. If a preemptable thread is waiting to enter the initial state it may be said that it is in the "pre-initial" state and hence it has a tick to complete. In this way there is no need for a data structure which has to distinguish between active and inactive threads.

8.2.2.2 States Visible in Specification

The code for each state in ft consists of one or more outgoing transitions. In Chap. 4, a state had to have an outgoing transition for every input combination for the purpose of analysing reactivity and determinism. Many of those transitions looped

```
ft_entry:    if FT_start(ft) is not resolved then
                 set pc to ft_entry
                 goto section pointed by next_thread
             else if FT_start(ft) is present then
                 set pc to ft_s1_t1 (highest priority transition of initial state s1)
                 emit and resolve FT_start signals for FTs that refine s1
                 set high ST_start_request signals for STs that refine s1
                 if s1 is a rendezvous state then
                     set high FT_rendezvous_ready(s1)
                 end if
                 goto ft tick end
             else
                 set pc to ft entry
                 goto ft tick end
             end if
ft_s1_t1:    if not all signals in the trigger are resolved then
                 set pc to th1_s1_t1
                 goto section pointed by next_thread
             else if trigger and variable conditions are true then
                 emit and resolve specification-visible local signals
                 set high specification-visible output signals
                 do procedures
                 determine next state and write it into next_state
                 emit FSM_start signals for FTs that refine next state
                 emit and resolve FSM_stop signals for FTs that refine this state
                 set high ST_start_request signals for STs that refine next state
                 set high ST_stop_request signals for STs that refine this state
                 if this state is a rendezvous state then
                     set low FT_rendezvous_ready signal for this state
                 end if
                 if next state is a rendezvous state then
                     set high FT_rendezvous_ready signal for next state
                 end if
                 goto ft exit
             end if
.
ft s1 tn1
.
.
ft sm tn1
```

Fig. 8.5 FSM thread

```
ft s_m t_nm
ft exit:      if FT_stop(ft) is not resolved then
                  set pc to ft exit
                  goto section pointed by next_thread
              else if FT_stop(ft) is present then
                  emit and resolve FSM_stop signals for lower level FTs
                  cancel FSM_start signals for lower level FTs
                  set high ST_stop_request signals for lower level STs
                  cancel ST_start_request signals for lower level STs
                  goto ft entry
              else
                  copy next_state to th1_pc
                  goto ft tick end
              end if
ft tick end:  resolve all unresolved signals
              set high the ft bit in threads_done
              goto tick handler
```

Fig. 8.5 (continued)

back to the source state without producing any outputs. Such transitions are not implemented. If a transition does not leave a state, it is implemented only if it produces outputs i.e. it emits signals or calls procedures. In Fig. 8.5 state s_1 has n1 transitions; state s_m has nm transitions. Each transition is realized with a two branch if-else construct. The if-else constructs are arranged in the order of transition priorities. The transition with the highest priority is listed first. In each transition, it is first checked whether the local signals in the transition trigger are resolved. A Rendezvous_done signal may also be a part of the trigger but, as for input signals, the resolution is not an issue. If not all signals are resolved, the next FT is selected by the scheduler but the current state of ft will have to be visited again since the local tick has not been completed. If all signals are resolved, the trigger and variable conditions are evaluated to determine whether the transition should be executed. If it is false the next transition is executed. Otherwise the transition outputs are produced. Specification-visible signals are emitted. Local signals have to be resolved in addition. It is possible to immediately resolve local signals due to the absence of strong abort in DFCharts. Procedures are also executed. If a shared variable is updated inside a procedure the corresponding Update_shared_variable signal must be set high. The next state is determined, but *pc* is not immediately set to the next state because it has to be checked in ft exit if ft has been pre-empted. If the next state is hierarchical, FT_start signals for FTs that refine it are emitted but they cannot be resolved immediately since they may be cancelled in ft1 exit if it turns out that ft has been pre-empted. FT_stop signals are emitted and resolved immediately for FTs that refine the current state. Start request signals are set high for STs that refine the next state while stop request signals are set high for STs that refine the current state. If the current state makes a transition back to itself, FT_stop and FT_start signals are emitted for the same FTs. The same applies for STs. The FT_rendezvous_ready signals are also handled if the current or next states are rendezvous. After all the outputs have been created the control jumps to ft exit.

8.2.2.3 Thread Exit

In ft exit it is checked by reading signal FT_stop(ft) whether ft has been pre-empted by the higher level thread. If the preemption has taken place the FT_stop signals for lower level FSM threads are emitted and resolved. This is necessary in case the current state is hierarchical and no transitions have been taken out of it. The same is done for lower level STs. Furthermore all emitted FT_start signals and ST_start_request signals are cancelled. The control jumps to ft entry, not to ft tick end, since ft may be restarted in the current tick. If the preemption has not taken place *pc* is simply set to the next state and ft tick end is executed next.

8.2.2.4 Local Tick End

In ft tick end, ft acknowledges that it has completed its local tick by asserting its bit in *threads_done* variable. Also, all signals that have not been resolved previously are resolved here. This is necessary since only emitted signals are resolved in code sections for transitions, not ones that are absent.

8.2.3 Hierarchical and Parallel Compositions

Figure 8.6 shows a DFCharts specification consisting of three FSMs. Without any FSM compositions, three FTs are needed for implementation. Figure 8.7 shows the flow of control among FTs, scheduler and tick handler for a single processor implementation. ft1, ft2 and ft3 implement FSM1, FSM2 and FSM3 respectively. Since ft3 is a preemptable FSM thread, it has to have thread entry and thread exit sections where it reads FT_start and FT_stop signals that are written by ft1. In general, these signals can be unresolved when they are read. For this reason, when the control reaches ft3 entry or ft3 exit it may immediately flow back to the scheduler. Similarly, local signal b can be unresolved when read by ft1. Thus the control may flow from ft1 s12 t1 to the scheduler. A schedule can take into account communication dependencies between FTs so that FT_start, FT_stop and b are always resolved when they are read. If ft1 is executed before ft3 in each global tick, FT_start and FT_stop will always be resolved when read. The same can be achieved for specification-visible signal b if ft2 is executed before ft1.

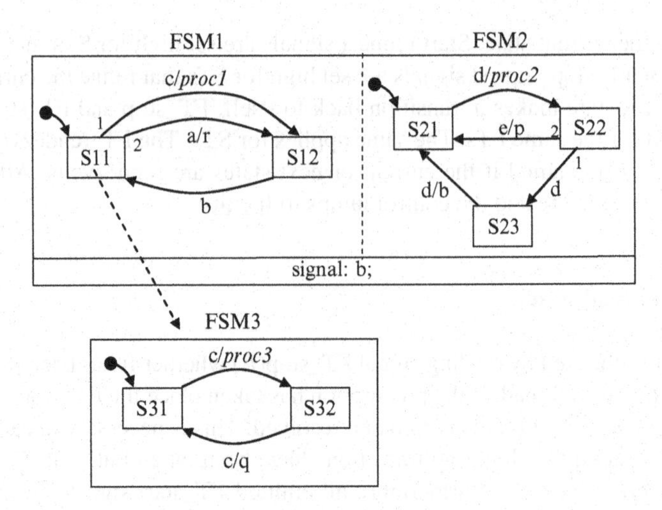

Fig. 8.6 A DFCharts specification consisting of three FSMs

The hierarchical composition of two FSMs follows the semantics of the refinement operator from Sect. 4.1.5. Figure 8.8 shows the implementation of the specification from Fig. 8.6, where ft1 implements the hierarchical composition of FSM1 and FSM3 while ft2 implements FSM2 as in Fig. 8.7. The FSM that represents the hierarchical composition of FSM1 and FSM3 has three states: S11S31, S11S32 and S12. FT_start and FT_stop signals are no longer needed. Thread entry and thread exit section from Fig. 8.7 also disappear. However, FSM1 and FSM3 cannot be executed concurrently by different processors. The potential benefit of executing *proc1* and *proc3* simultaneously would no longer be available.

The parallel composition of two FSMs follows the semantics of the synchronous parallel operator from Sect. 4.1.2. The states of the resulting FSM are created by taking the cross product of the states that belong to the input FSM threads. As with the hierarchical composition, internal signals are removed, but concurrency disappears as well. The additional concern with the parallel combination is the potential state explosion that results from the cross product. For example, if FSM1 and FSM2 from Fig. 8.6 are merged, local signal b is no longer needed and the resulting FSM has six states: S11S21, S11S22, S11S23, S12S21, S12S22 and S12S23.

In the DFCharts automata semantics the operators are applied to create a single FSM, which represents the behaviour of a whole specification. The equivalent FSM is constructed bottom-up, by starting from the lowest hierarchical levels and moving upwards to the top level. For example, in Fig. 8.6, FSM1 and FSM3 would have to be merged first. The result is then combined with FSM2. This restriction does not apply here. FSM1 and FSM2 can be merged by the parallel composition while leaving FSM3. Before the parallel composition, FSM3 refines state S11. After the parallel composition, FSM3 refines states S11S21, S11S22 and S11S23.

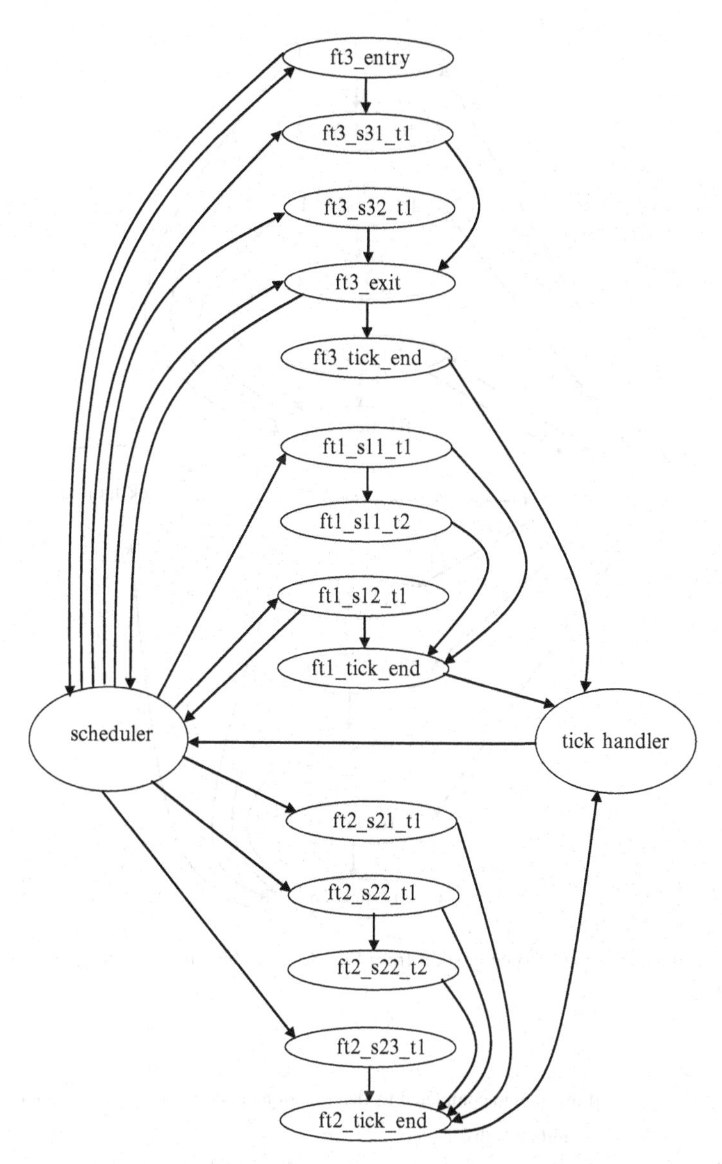

Fig. 8.7 Implementation of specification from Fig. 8.6 without any FSM compositions

8.2.4 FSM Scheduler

The template for the FSM scheduler is shown in Fig. 8.9. It specifies the order in which threads are run. The schedule is static; threads are run in the same order in every processor tick. After the system's start-up the processor begins execution at run_ft$_1$. In each processor tick, the schedule is repeated until all

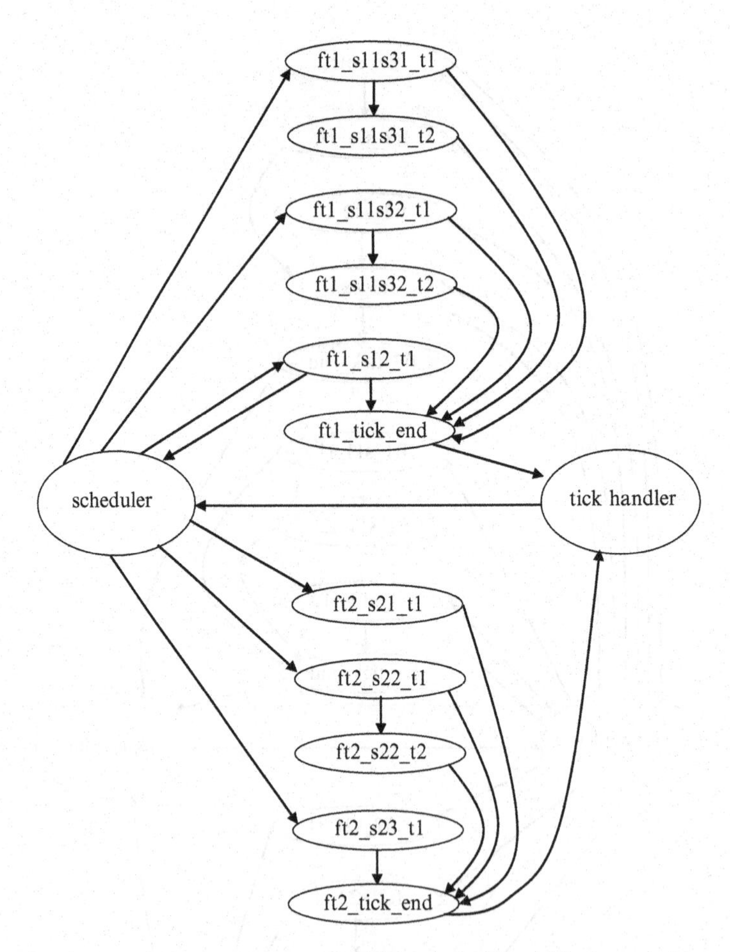

Fig. 8.8 Implementation of specification from Fig. 8.6 with hierarchical composition of FSM1 and FSM3

run ft_1: **if** this thread completed its tick (its bit in *threads_done* is high) **then**

  goto run th_2

else

  set *next_thread* to run th_2

  goto location contained in *pc* of ft_1

end if

run ft_2

.

run ft_m

Fig. 8.9 FSM scheduler

threads have completed their local ticks. If a thread has already completed its local tick, it is skipped.

The order in which threads are executed can have a large impact on the system performance. A fixed execution order in every global tick will not be ideal for many applications, since data dependencies among threads may not be static. On the other hand, finding an optimal schedule could be greatly complicated by a distribution of threads on multiple processors. This could be a topic for further investigation.

8.2.5 Master Tick Handler

The tick handler is different for master and slave processors. We first consider the tick handler for the master processor. The template describing its function is shown in Fig. 8.10. The point in the code marked by tick handler is arrived at when an FT finishes its local tick. *thread_done* variable indicates by its individual bits which threads have finished their local ticks. If all bits in *thread_done* are not high, the processor tick end has not yet been reached. The control moves to the scheduler which runs the next thread. Otherwise, the next task is to perform actions at the tick end.

The master processor first has to wait for all slave processors to complete their ticks. When the Tick_finished signal has been set high by every slave processor, the global tick is completed. At this point the master tick handler can deal with signals related to rendezvous channels and STs. However, it may have to wait first for one or more STs to complete their iterations. This is discussed later in this section. All Rendezvous_done signals from the previous tick are cleared. Then, the Rendezvous_ready signals from both sides (FT and ST) are checked for each rendezvous channel. If both sides are ready data is transferred between the shared memories and Rendezvous_done is set high. ST_stop_request and ST_start_request signals are checked for each ST. If ST_stop_request is high, ST_stop signal is set high. ST_start_request causes the same action on ST_start signal. It should be emphasized that, when both ST_stop_request and ST_starts_request signals are high, the ST must first be issued a stop command by ST_stop and then a start command by ST_start. The opposite order is never valid.

While the master processor was doing rendezvous and ST related operations in the tick handler section, slave processors were waiting for the Start_tick_end_actions signal. The master processor sends this command once it has finished with rendezvous channels and STs. The command tells each slave processor to update shared variables that are shared by its FTs. When updating is done, a slave processor has to clear all signals that were written in the previous tick. The master processor also has to update shared variables that are shared by its FTs, but , in addition, it also has to update shared variables that are shared by FTs executed on different processors. A slave processor acknowledges that it has finished updating shared variables and clearing signals by

setting high Tick_end_actions_completed. When the master processor receives the acknowledgment from all slave processors, it clears all its signals except for Rendezvous_done signals. Then, it starts a new global tick by making a short pulse Start_tick signal.

```
tick_handler:  if all FTs finished their ticks (all bits in threads_done high) then
                   for i in 1 to _number_of_processors do
                       await Tick_finished(i)
                   end for
                   wait for SDF iterations to make required number of ticks (optional)
                   for i in 1 to _number_of_rendezvous_channels do
                       clear Rendezvous_done(i)
                       if FT_rendezvous_ready(i) and ST_rendezvous_ready(i)
                               are both present then
                           do data transfer and emit Rendezvous_done(i)
                       end if
                   end for
                   for i in 1 to _number_of_ST_threads do
                       if ST_stop_request(i) is present then
                           set high ST_stop(i)
                       end if
                       if ST_start_request(i) is present then
                           set high ST_start(i)
                       end if
                   end for
                   clear threads_done
                   set high Start_tick_end_actions for all slave processors
                   update shared variables
                   for i in 1 to _number_of_slave_processors do
                       await Tick_end_actions_completed(i)
                   end for
                   clear all signals except for Rendezvous_done signals
                   emit Start_tick to all slave processors
                   goto run th1
               else
                   goto section pointed by next_thread
               end if
```

Fig. 8.10 Master tick handler

Before dealing with rendezvous channels and other tick-end activities, the master tick handler may have to ensure that the current iteration of an SDF graph completes the specified number of ticks. The influence of SDFG speeds on the system behaviour was discussed in Sect. 4.3. Only SDF graphs that communicate with FSMs have to be taken care of.

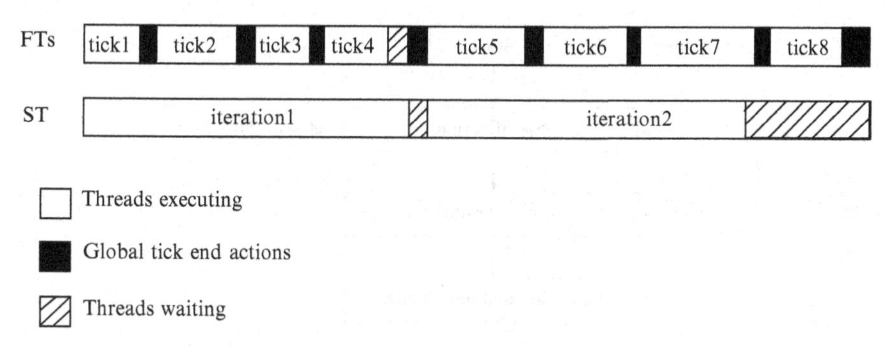

Fig. 8.11 Control of SDF iteration length

Figure 8.11 illustrates the action of the master tick handler needed to control the length of SDF iterations. In this example, we assume that it has been specified that each iteration of an SDF graph takes four ticks to complete. At the end of the fourth tick the master tick handler finds that the SDF thread has still not finished the current iteration. Consequently, it has to wait for the end of the iteration before starting tick end actions. Thus, the first iteration effectively stretches the fourth tick. The other form of control is seen in the next four ticks that are longer than the first four. The ST completes its iteration during the seventh tick and becomes blocked waiting for rendezvous. At the end of the seventh tick, the master tick handler ignores the rendezvous ready signal of the ST since the iteration was completed too early with respect to the number of ticks elapsed in the FSM domain. The rendezvous will happen at the end of the eighth tick.

The actions performed by the master tick handler in Fig. 8.11 are necessary to guarantee that the system will have exactly single and unique behaviour. When the speed of an SDFG is not controlled, the system behaviour may be sensitive to the relative speeds of the processing elements, as discussed in Sect. 4.3. In that case, it may be necessary to perform HW/SW co-simulation to ensure that the implementation behaves as intended. This is shown in Fig. 8.12. At this level of abstraction, verification is much more difficult. For safety critical applications, where the correctness of the final implementation is extremely important, controlling SDFG speeds is preferable. It should be emphasized, though, that the operations that control SDFG speeds create significant overhead, which results in higher implementation cost. For this reason, the design flow in Fig. 8.12 could be more suitable for applications that are highly cost-sensitive and not safety-critical such as various consumer electronics products.

When each SDF iteration takes a certain and fixed number of ticks of the FSM clock, the implementation resembles a multirate synchronous system. Distributed implementation of synchronous specifications created in Esterel [15], Lustre [16] and Signal [17] has been researched extensively. Unlike DFCharts, synchronous languages do not provide an easy trade-off between implementation efficiency and verification effort. In DFCharts design flow, an SDFG may be forced to follow the FSM clock, but this is not compulsory. Communication between SDFGs and FSMs can be asynchronous, which is likely to result in lower implementation cost. On the

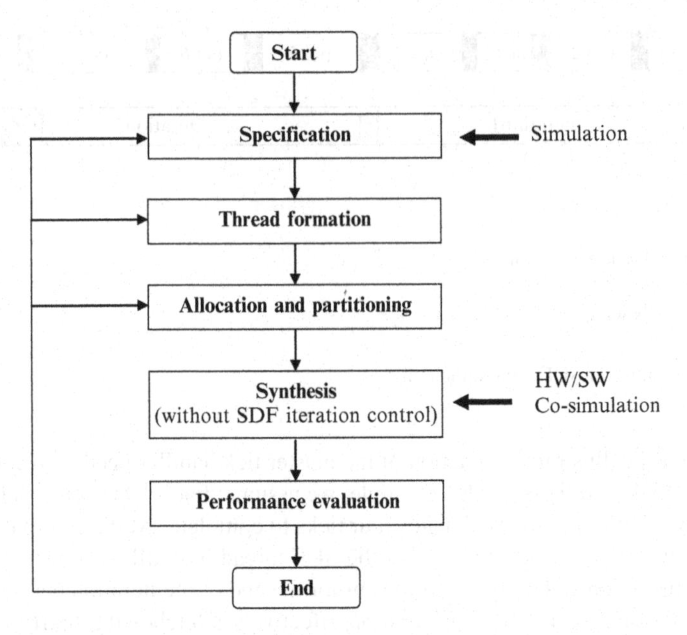

Fig. 8.12 DFCharts design flow without SDF iteration control

other hand, synchronous languages do not support asynchronous communication mechanisms. Signal [17] offers some support for asynchrony, since it allows unrelated clocks, but creating asynchronous communication mechanisms like rendezvous and FIFO buffer is very difficult in Signal.

8.2.6 Slave Tick Handler

The slave tick handler is simpler than the master tick handler as can be seen from the template in Fig. 8.13.

The previous section has showed that the master tick handler has five functions: controlling the length of SDF iterations, moving data across rendezvous channels, starting and stopping STs, updating shared variables and clearing signals. The slave tick handler only has to update local shared variables and clear signals.

8.3 Frequency Relay Implementation

The specification for the frequency relay consisting of seven FSMs and one SDFG was presented in Sect. 3.2. Each iteration of SDF1 is set to take five FSM ticks, since FSM3 takes five ticks on its path that leads back to the initial state S31, where it is ready again to receive data from SDF1.

```
tick_handler:  if all FTs finished their ticks (all bits in threads_done high) then
                   emit Tick_finished
                   await Start_tick_end_actions
                   update shared variables
                   emit Tick_end_actions_completed
                   await Start_tick
                   clear all signals
                   goto run th1
               else
                   goto section pointed by next_thread
               end if
```

Fig. 8.13 Slave tick handler

Among various ways in which FSMs from the frequency relay can be merged, the hierarchical composition of FSM2 (Fig. 3.9) and FSM7 (Fig. 3.10) is the most useful one. When a single thread is used for implementation of FSM2 and FSM7 instead of two threads, specification-visible signal alld, specification-invisible signals FT_start and FT_stop, and code sections tick exit and tick entry all disappear. At the same time, there is no significant loss of concurrency. *update()*, which is the longest procedure in FSM2, cannot be invoked while FSM7 is active. The outgoing transitions of S22, which is refined by FSM7, do not call any procedures. Therefore, there would not be any performance gain if FSM2 and FSM7 were implemented with two threads running on different processors. Other FSM compositions are less attractive, since they offer both advantages and disadvantages that have to be weighted carefully. For example, if FSM5 and FSM6 (Fig. 3.8) are merged, specification-visible signal st disappears but the resulting FSM has eight states (4×2) which is more than the total of six states for FSM5 and FSM6. In the following steps, we will only use the hierarchical composition of FSM2 and FSM7. Thus, there will be six FSMs to partition and allocate: FSM1, FSM27, FSM3, FSM4, FSM5 and FSM6.

There are many allocation and partitioning options for the frequency relay. We limit our attention to two, so that we can describe them in more detail. SDF1 has to perform intensive calculations involving long loops, which could not be done with a software implementation within the sampling period of the AC signal (125 μsec). Thus, SDF1 is implemented with a hardware functional unit in both options. FSMs are implemented with one ReMIC in the first option, and two ReMICs in the second option. The partitioning for the second option is shown in Fig. 8.14. Although the master processor runs a smaller number of FSMs, its execution load is usually not smaller than that of the slave processor, since FSM27 often invokes the lengthy procedure for updating the thresholds. The other reason for this partitioning is that the two groups of FSMs do not communicate by any specification-visible local signals.

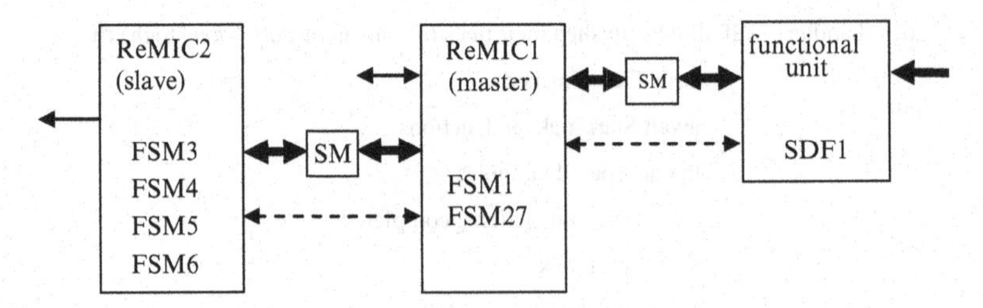

Fig. 8.14 Partitioning for the second implementation option

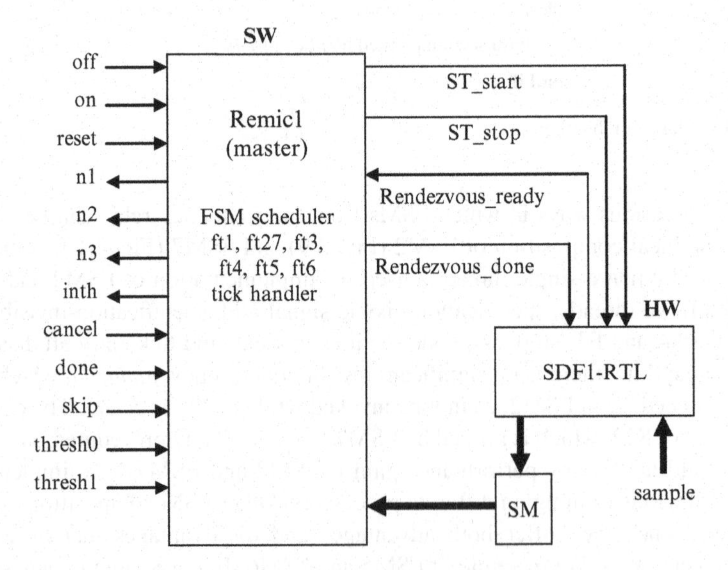

Fig. 8.15 Frequency relay implementation with one ReMIC processor

Figures 8.15 and 8.16 show the two implementation options after the synthesis step. Note that only the logical flow of data is showed between shared memories. Address and data buses will depend on the implementation technology. The signals labelled n1, n2 and n3 are the switches for controlling the network load. In Fig. 8.16 the resolve lines are marked with *r* as in FT_stop_r. Schedules were made taking into account data dependencies between the FTs. The schedule for the master processor in the first implementation is ft1, ft27, ft3, ft4, ft5, ft6. The schedule for the master processor in the second implementation is ft1, ft27 while the schedule for the slave processor is ft3, ft4, ft5, ft6.

The lengths of the threads for the first implementation in terms of the number of ReMIC instructions are given in Table 8.1. The numbers are identical for the second implementation except for ft1 which has 20 more instructions due to additional synchronization in the system.

The program and data memory requirements are shown in Table 8.2. For the second solution, the numbers from individual processors (seen in the brackets where

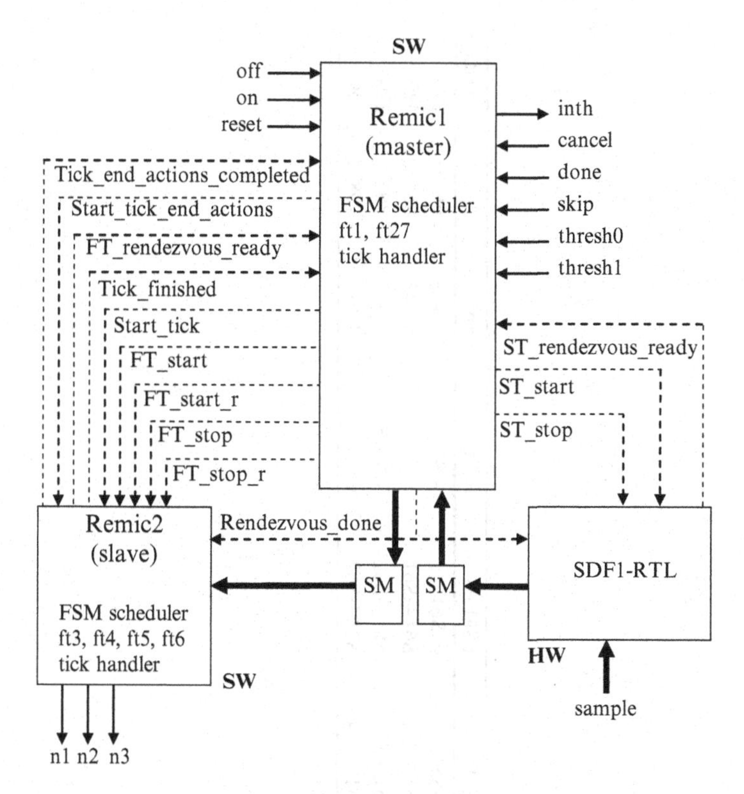

Fig. 8.16 Frequency relay implementation with two ReMIC processors

Table 8.1 Thread lengths in the first solution

Thread	ft1	ft27	ft3	ft4	ft5	ft6
Instructions	73	208	131	119	278	88

Table 8.2 Program and data memories

	Program memory (bytes)	Data memory (bytes)
1p	4,296	118
2p	4,584 (1,776+2,808)	122 (42+80)

the first number represents the master processor) are added up. The combined data memory is slightly larger for the second solution due to duplication of variables *threads_done* and *next_thread*. The program memory is also larger due to increased code size for tick handlers.

Table 8.3 presents a performance comparison between the two implementation options. Four rows represent four global ticks. FSM transitions are shown by giving source and sink states. The source state is omitted if the FSM has just become active

Table 8.3 Performance comparisons between two implementations

FSM1	FSM2	FSM3	FSM4	FSM5	FSM6	FSM7	1p	2p	speed up
S11→S12	→S21	→S31	→S41	→S51	→S61	→S71	933	666	29%
S12→S12	S23→S21[a]	S33→S34[b]	S41→S42	S51→S51	S61→S61	X	1,302	885	32%
S12→S12	S22→S22	S32→S33	S41→S41	S51→S51	S61→S61	S74→S75[c]	1,047	801	23%
S12→S12	S23→S21[d]	S34→S31	S43→S41	S51→S54	S61→S62	x	1,146	798	30%

[a] Six thresholds updated
[b] The value of *fs* is zero
[c] Transition 3 taken
[d] Four thresholds update

in the current tick. Additional notes are included where applicable. They indicate exactly which transition is taken when there are multiple transitions between two states, or the value of a variable, since that can also influence the amount of time consumed. Columns 8 and 9 show the time taken by the first and the second implementation, respectively, measured in ReMIC cycles. The last column shows the percentage difference. Clearly, there is a performance/cost trade-off between the two implementations.

As mentioned previously FSM3 takes five ticks on its path from S31 to S31. There are four transitions on the path from S31 to S31, but FSM3 has to spend two ticks in S34 as it waits for FSM4 to make the transition from S42 to S43. Calculating exactly the maximum number of ReMIC instruction cycles that these five ticks can take is required to show whether SDF1 will always be ready to receive the input on ch1 in time. However, this task is quite intensive since there are many possible execution paths in the system. The result does not only depend on FSM3 and FSM4 but on the other FSMs as well, since they all execute in lock-step and hence influence the tick duration. Currently, all FSMs in a DFCharts specification must use the same clock, but this could change in a future development. Apart from the states of the FSMs, values of variables can also have an effect, since they can influence computation time inside procedures.

A tool would be needed to provide an accurate, automated analysis. However, a designer can still produce quick, conservative estimates of the initial performance requirements. The longest tick in the system is shown in row 2 of Table 8.3. In that tick, FSM2 updates all six thresholds, FSM3 calculates the frequency status, and FSM4 calculates the average rate of change. For the single processor implementation, the tick lasts 1302 ReMIC instructions. Multiplying this number by five gives 6,510. With two processors, this would be reduced to $5 \cdot 885 = 4{,}425$. Finally, the sampling frequency needs to be taken into account. All instructions that comprise the five ticks must be completed within the period of the sampling clock. Otherwise, SDF1 would miss the input on ch1 while waiting for FSM3 to reach S1. For the sampling frequency of 8 kHz, the minimum clock frequency of the single ReMIC is 8 kHz $\cdot$ 6,510=52.08 MHz. For the two processor implementation the minimum clock frequency is 8 kHz $\cdot$ 4,425=35.4 MHz.

Of course, these numbers are largely overestimated. The five ticks under consideration could not all be of the maximum duration. However, the results above do provide a useful initial boundary on the processor performance. As for the hardware that implements SDF1, it needs to be able to complete an iteration is slightly less then 1/fs, where *fs* is the sampling frequency.

Chapter 9
Conclusions

9.1 Summary and Overview

Design of embedded systems based on formal models of computation has been gaining acceptance as a sound method for dealing with increasing system complexities. While models of computation have been successfully used individually for control-dominated systems and data-dominated systems, modelling of heterogeneous systems still poses a challenge. We presented DFCharts, a model of computation for heterogeneous embedded systems that combine control-dominated and data-dominated parts. We demonstrated how DFCharts could be used in embedded systems design by linking it to system level design languages and implementation architecture. We used a realistic heterogeneous embedded system called frequency relay, in order to illustrate the concepts.

DFCharts integrates the hierarchical concurrent finite state machines (HCFSM) model with synchronous dataflow (SDF). As in Argos, FSMs are composed using synchronous parallel, hiding, and refinement operators. The fourth operator called asynchronous parallel is used to connect FSMs with SDFGs. The refinement operator also allows the state of an FSM to be refined by an SDFG. The SDFG becomes active when the state is entered, and it gets instantly terminated when the state is left. Due to the application of the asynchronous parallel and refinement operators, behavioural heterogeneity is addressed using both hierarchical and parallel compositions. This contributes to modelling flexibility, which is important for capturing the behaviour of a system accurately and producing efficient implementations. The asynchronous parallel operator has been designed so that FSM and SDF, two vastly different models, retain their original characteristics as much as possible. Towards this aim, it allows an SDFG to operate at its own speed. An SDFG only has to synchronize with FSMs between two iterations when it is receiving inputs for the next iteration and sending outputs produced during the previous iteration. Besides SDF, any dataflow model with clearly defined iterations and bounded memory can easily be incorporated. Thus, DFCharts is expressive enough to cover a wide range of embedded systems.

I. Radojevic and Z. Salcic, *Embedded Systems Design Based on Formal Models of Computation*, DOI 10.1007/978-94-007-1594-3_9,

The formal semantics of DFCharts was presented in Chap. 4. We described the ordering of events on a single hierarchical level of DFCharts using the tagged signal model (TSM) framework. The TSM semantics of DFCharts closely focuses on data transfer between FSMs and SDGFs. It shows how values in an FSM array variable appear as multiple tokens in an SDFG buffer, when data is transferred across a rendezvous channel. The automata based semantics, which resembles the semantics of Argos, can be used for the global analysis of a DFCharts specification. It expresses the behaviour of a DFCharts specification in terms of a single, flat FSM. This is achieved by representing the operation of each SDFG as an FSM. The abstract 'SDF FSMs' can then be merged with the 'real FSMs'. 'SDF FSMs' are significant in the automata semantics, since they allow a single formalism to embrace both FSMs and SDFGs. When a single FSM is obtained, determinism and reactivity can be analysed. Only specifications that satisfy these two properties are correct. In the automata semantics, data transfer across rendezvous channels is not analysed. All that matters is the event that is generated when a rendezvous occurs. Thus, the automata semantics and TSM semantics complement each other – the former looks at the behaviour of a complete system, the latter focuses on communication through rendezvous channels. An important feature of the automata semantics is the use of multiclock FSMs. In a multiclock FSM, transitions are triggered by different clocks. Apart from DFCharts, this concept can be used for describing the semantics of other models where multiple clocks appear.

DFCharts was used in Chap. 5 to analyse the ability of SystemC and Esterel to specify heterogeneous embedded systems. We examined the level of support that the two languages provide for features necessary to capture the behaviour of heterogeneous embedded systems and are found in DFCharts, such as synchronous events, rendezvous channels, FIFO channels, hierarchy, preemption etc. Some features are completely supported, others are more difficult to describe. Possible modifications were suggested for both languages in their weak areas. For SystemC, it is mainly control-dominated behaviour. For Esterel, the focus is on data-dominated behaviour, as could have been expected.

In general, it takes a lot of effort to verify the correctness of a multi-threaded Java design [8] due to deadlocks that are difficult to detect. It may happen that a design performs correctly before it suddenly crashes for no apparent reason. The Java class library described in Chap. 6 provides an opportunity for a more reliable design in Java. It contains classes that enable making specifications according to the DFCharts model. Instead of using threads with mutexes, locks, and other mechanisms commonly employed in Java that often lead to unpredictable behaviour, the designer specifies FSMs and SDFGs first, and , then, connects them with synchronous signals and rendezvous channels. The result is a design with clear semantics.

Chapter 7 reviewed contemporary trends in multiprocessing architectures and then proposed HETRA, a multiprocessor architecture that has special features for both control-dominated and data-dominated behaviours unlike most other architectures which mainly concentrate on data processing and achieving high throughput. HETRA's specific implementation, called HiDRA, based on multiple reactive processor cores, is particularly suitable for DFCharts implementation.

In Chap. 8, we presented a design methodology with a complete design flow from specification to implementation. DFCharts is used for specification, while HiDRA (a subset of HETRA) is used for implementation. An important strength of the methodology is that it starts by capturing the behaviour of a system without any reference to implementation. Besides fast verification, this also allows the designer to easily explore various mapping options before the SW/HW synthesis. We laid a foundation for automated synthesis of heterogeneous embedded systems by specifying in detail how DFCharts is executed on HiDRA. Automated synthesis is a key feature that is missing from most system-level design methodologies. It provides a much bigger improvement in design productivity than just raising the initial level of abstraction above the SW/HW boundary and then using manual refinement to obtain a final implementation. A necessary condition for automated synthesis is to use a model of computation with precisely defined semantics. For this reason the design methodology in Chap. 8 is based around DFCharts. The frequency relay case study was used to show the practical application of the methodology. Two implementation options were presented, which demonstrated the trade-off between performance and cost.

9.2 Future Research

9.2.1 DDFCharts Design Flow

Since DDFCharts is based on DFCharts, the DDFCharts design flow would have many similarities to that of DFCharts. However, it would be significantly more complex in some aspects. The implementation architecture of DDFCharts would likely consist of many more parallel processors. This would make design space exploration more difficult. One area in particular that deserves special attention is the estimation of worst case execution time (WCET). For hard real-time embedded systems WCET is essential. However, WCET estimation in massively parallel systems is a very difficult task.

9.2.2 Hardware Implementation

The current DFCharts based design flow finishes with software implementation on a multiprocessor architecture. Hardware synthesis of DFCharts models would be a useful implementation option. Apart from rendezvous, which appears in asynchronous parallel operators, all other computation and communication mechanisms in DFCharts would have more or less straightforward implementation in hardware. According to DFCharts semantics, when rendezvous occurs, the ticks of the two clocks involved coincide. This is achievable in DFCharts software implementation, where a clock is just a logical concept and clock ticks are not related to physical time. In digital hardware circuits, clock ticks are clearly defined as clock periods whose timing is controlled by an oscillator. It would be very difficult to synchronize two different oscillators at certain points in time. Hence, the DFCharts semantics could not be literally followed in hardware and would have to be reinterpreted somewhat. For example, each clock tick in a DFCharts model could be translated into multiple ticks of the physical clock. This would allow more flexibility in the implementation of rendezvous.

9.2.3 Strong Abort

As in Argos, only weak abort is available in DFCharts. Strong abort, where an FSM is not allowed to produce outputs in the instant of preemption, would be a useful addition for modelling control-dominated behaviour. However, it would make handling of rendezvous more difficult. In the current semantics, when two rendezvous states associated with the same channel are reached, the rendezvous has to happen. With strong abort, this assumption would have to be lifted leading to more complicated semantics and implementation.

9.2.4 Including More Expressive Dataflow Models

Inputs arrive periodically in most signal process applications. Consequently, data rates are static on all internal and external channels. For such applications, SDF and related static dataflow models can produce very efficient implementations. On the other hand, there are a significant number of applications with variable data rates, especially in the multimedia domain. In DFCharts, variability in data rates can be handled to some extent by using several different SDF graphs at run time. However, it would certainly be useful to include a model with dynamic dataflow like Kahn process networks (KPN). Unlike SDF, KPN does not have an iteration. The main issue would be when to exchange data between a KPN and FSMs.

9.2.5 Program State

Imperative statements inside a state that can be compiled into an FSM would be a useful addition to the graphical syntax of DFCharts. This could be done in a similar fashion as in Synccharts where Esterel programs can be inserted in states. Instead of Esterel, DFCharts could use SystemJ [114], which is built on top of Java. A DFCharts state could be refined by a SystemJ *reaction*, which is comparable to a module in Esterel. Only reactions that can be compiled into an FSM with datapath should be placed in a DFCharts specification. Dynamic memory creation should not be allowed. This requirement stems from the semantics of DFCharts. Further research on how to describe SDF and related dataflow models in SystemJ would open up a possibility of having the complete DFCharts model captured in SystemJ.

9.2.6 *Formal Verification*

While formal verification for DFCharts has not yet been developed, the formal semantics of DFCharts based on MCFSM represents a large step towards this goal. MCFSM can be used as an efficient input model for formal verification. We demonstrated in Chap. 4 that the MCFSM model captures a mixed synchronous/asynchronous system with fewer transitions than the common approach of using a fictictitious base clock to read real clocks. However, parallel products in MCFSM can create state explosion as in other models. State explosion may adversely impact the scalability of DFCharts, since it makes verification of complex systems increasingly difficult. Therefore, this issue requires a careful investigation.

9.2.7 *Proof of Correctness for DFCharts Implementation*

Chapter 8 provides detailed description of how an implementation is produced from a DFCharts specification. The methodology would be further strengthened by providing a proof of equivalence between specification and implementation levels.

References

1. D. Edenfeld, A.B. Kahng, M. Rodgers, Y. Zorian, 2003 technology roadmap for semiconductors. IEEE Comput. **37**(1), 47–56 (2004)
2. S. Edwards, L. Lavagno, E.A. Lee, A. Sangiovanni-Vincentelli, Design of embedded systems: formal methods, validation and synthesis. Proc. IEEE **85**(3), 366–390 (1997)
3. S. Edwards, *Languages for Digital Embedded Systems* (Kluwer, Dordrecht/Boston, 2000)
4. E. Clarke, O. Grumberg, D. Peled, *Model Checking* (MIT Press, Cambridge, 1999)
5. A. Jantsch, I. Sander, Models of computation and languages for embedded system design. IEE Proc. Comput. Digit. Technol. **152**(2), 114–129 (2005)
6. E.A. Lee, S. Neuendorffer, Concurrent models of computation for embedded software. IEE Proc. Comput. Digit. Technol. **152**(2), 239–250 (2005)
7. J. Hopcroft, J. Ullman, *Introduction to Automata Theory, Languages, and Computation* (Addison-Wesley Publishing Company, Reading, 1979)
8. E.A. Lee, The problem with threads. IEEE Comput. **39**(5), 33–42 (2006)
9. C.G. Cassandras, *Introduction to Discrete Event Systems* (Kluwer, Dordrecht/Boston, 1999)
10. E.A. Lee, T.M. Parks, Dataflow process networks. Proc. IEEE **83**, 773–801 (1995)
11. A. Benveniste, G. Berry, The synchronous approach to reactive and real-time systems. Proc. IEEE **79**(9), 1270–1282 (1991)
12. T. Murata, Petri nets: properties, analysis, and applications. Proc. IEEE **77**(4), 541–580 (1989)
13. C.A.R. Hoare, Communicating sequential processes. Commun. ACM **21**(8), 666–677 (1978)
14. R. Milner, *Communication and Concurrency* (Prentice-Hall, Englewood Cliffs, 1989)
15. G. Berry, G. Gonthier, The Esterel synchronous programming language: design, semantics, implementation. Sci. Comput. Program. **19**(2), 87–152 (1992)
16. N. Halbwachs, P. Caspi, P. Raymond, D. Pilaud, The synchronous data flow programming language LUSTRE. Proc. IEEE **79**(9), 1305–1320 (1991)
17. P. Le Guernic, T. Gautier, M. Le Borgne, C. Le Maire, Programming real-time applications with SIGNAL. Proc. IEEE **79**(9), 1321–1336 (1991)
18. www.esterel-tecnologies.com
19. F. Balarin et al., *Hardware-Software Co-Design of Embedded Systems: The Polis Approach* (Kluwer, Boston/Dordrecht, 1997)
20. J. Eker et al., Taming heterogeneity – the ptolemy approach. Proc. IEEE **91**(1), 127–144 (2003)
21. www.mathworks.com
22. D. Harel, Statecharts: a visual formalism for complex systems. Sci. Comput. Program. **8**(3), 231–274 (1987)
23. E.A. Lee, D.G. Messerschmitt, Synchronous data flow. Proc. IEEE **75**(9), 1235–1245 (1987)

I. Radojevic and Z. Salcic, *Embedded Systems Design Based on Formal Models of Computation*, DOI 10.1007/978-94-007-1594-3,

24. F. Maraninchi, Y. Remond, Argos: an automaton-based synchronous language. Comput. Lang. **27**(1–3), 61–92 (2001)
25. E.A. Lee, A. Sangiovanni-Vincentelli, A framework for comparing models of computation. IEEE Trans. Comput. Aided Des. Circ. Syst. **17**(12), 1217–1229 (1998)
26. G. Bilsen, M. Engels, R. Lauwereins, J.A. Peperstraete, Cyclo-static dataflow. IEEE Trans. Signal Process. **44**(2), 397–408 (1996)
27. G. Kahn, The semantics of a simple language for parallel programming, in *Proceedings of IFIP Congress 1974*, Stockholm, Aug 1974, pp. 471–475
28. I. Radojevic, Z. Salcic, P. Roop, A new model foe heterogeneous embedded systems: what Esterel and SyncCharts need to become a suitable specification platform. Int. J. Softw. Eng. Knowl. Eng. **15**(2) (2005)
29. B.A. Davey, H.A. Priestley, *Introduction to Lattices and Order* (Cambridge University Press, Cambridge, 1990)
30. T.M. Parks, Bounded scheduling of process networks. Ph.D. dissertation, Technical Report UCB/ERL 95/105, Department of EECS, University of California, Berkeley, 1995
31. G. Kahn, D.B. MacQueen, Coroutines and networks of parallel processes, in *Proceedings of the IFIP Congress 1977*, North-Holland, Aug 1977, pp. 993–998
32. E.A. Lee, D.G. Messerschmitt, Static scheduling of synchronous data flow programs for digital signal processing. IEEE Trans. Comput. **36**(1), 24–35 (1987)
33. E.A. Lee, Consistency in dataflow graphs. IEEE Trans. Parallel Distrib. Syst. **2**(2), 223–235 (1991)
34. J.T. Buck, Scheduling dynamic dataflow graphs with bounded memory using the token flow model. Ph.D. dissertation, Technical Report UCB/ERL 93/69, Department of EECS, University of California, Berkeley 1993
35. B. Bhattacharya, S. Bhattacharyya, Parameterized dataflow modeling for DSP systems. IEEE Trans. Signal Process. **49**(10), 2408–2421 (2001)
36. P.K. Murthy, E.A. Lee, Multidimensional synchronous dataflow. IEEE Trans. Signal Process. **50**(8), 3306–3309 (2002)
37. C. Park, J.W. Chung, S. Ha, Extended synchronous dataflow for efficient DSP system prototyping, in *Proceedings on Workshop in Rapid System Prototyping*, Clearwater, June 1999
38. H. Oh, N. Dutt, S. Ha, Memory optimal single appearance schedule with dynamic loop count for synchronous dataflow graphs, in *Proceedings of Asia and South Pacific Design Automation Conference (ASP-DAC'06)*, Yokohama City, Jan 2006
39. S. Stuijk, M. Geilen, T. Basten, Exploring trade-offs in buffer requirements and throughput constraints for synchronous dataflow graphs, in *Proceedings of Design Automation Conference (DAC '06)*, San Francisco, July 2006
40. S. Stuijk, M. Geilen, T. Basten, Minimising buffer requirements of synchronous dataflow graphs with model checking, in *Proceedings of Design Automation Conference (DAC'05)*, New York, June 2005
41. D. Bjorklund, Efficient code synthesis from synchronous dataflow graphs, in *Proceedings of Formal Methods and Models for Co-Design (MEMOCODE'04)*, June 2004
42. P.K. Murthy, S.S. Bhattacharyya, Shared buffer implementations of signal processing systems using lifetime analysis techniques. IEEE Trans. Comput. Aided Des. Integr. Circuits Syst. **20**(2), 3306–3309 (2001)
43. M. Ade, R. Lauwereins, J.A. Peperstraete, Data memory minimisation for synchronous data flow graphs emulated on DSP-FPGA targets, in *Proceedings of Design Automation Conference (DAC'97)*, Anaheim, June 1997
44. R. Govindarajan, G.R. Gao, P. Desai, Minimizing buffer requirements under rate-optimal schedule in regular dataflow networks, J. VLSI Sig. Proc. **31**(3), 207–229 (2002)
45. J. S. Kin, J.L Pino, Multithreaded synchronous data flow simulation, in *Proceedings of Design, Automation and Test in Europe Conference (DATE'03)*, Mar 2003
46. C. Hsu, S. Ramasubbu, M. Ko, J.L. Pino, S.S. Bhattacharyya, Efficient simulation of critical synchronous dataflow graphs, in *Proceedings of Design Automation Conference (DAC '06)*, July 2006

47. E. Zitzler, J. Teich, S.S. Bhattclcharyya, Evolutionary algorithms for the synthesis of embedded software. IEEE Trans. Very Large Scale Integr. Syst. **8**(4) (2000)
48. W. Sung, S. Ha, Memory efficient software synthesis with mixed coding style from dataflow graphs. IEEE Trans. Very Large Scale Integr. Syst. **8**(5), 522–526 (2000)
49. M. Sen, S.S. Bhattacharyya, Systematic exploitation of data parallelism in hardware synthesis of DSP applications, in *Proceedings of International Conference on Acoustics, Speech, and Signal Processing (ICASSP '04)*, May 2004
50. H. Jung, K. Lee, S. Ha, Efficient hardware controller synthesis for synchronous dataflow graph in system level design. IEEE Trans. Very Large Scale Integr. Syst. **10**(2), 672–679 (2002)
51. M.C. Williamson, E.A. Lee, Synthesis of parallel hardware implementations from synchronous dataflow graph specifications, in *Proceedings of Conference on Signals, Systems and Computers*, Nov 1996
52. A. Kalavade, P.A. Subrahmanyam, Hardware/software partitioning for multifunction systems. IEEE Trans. Comput. Aided Des. Integr. Circuits Syst. **17**(9), 819–837 (1998)
53. A. Kalavade, E.A. Lee, A hardware-software codesign methodology for DSP applications. IEEE Des. Test Comput. **10**(3), 16–28 (1993)
54. T. Wiangtong, P. Cheung, L. Luk, Hardware/software codesign: a systematic approach targeting data-intensive applications. IEEE Signal Process Mag. **22**(3), 14–22 (2005)
55. N. Halbwachs, *Synchronous Programming of Reactive Systems* (Kluwer, Dordrecht/Boston, 1993)
56. R. Budde, G.M. Pinna, A. Poigne, Coordination of synchronous programs, in *Proceedings of International Conference on Coordination Languages and Models, LNCS 1594*, Apr 1999
57. J. Colaco, B. Pagano, M. Pouzet, Specification and semantics: a conservative extension of synchronous data-flow with state machines, in *Proceedings of the 5th ACM International Conference on Embedded Software (EMSOFT '05)*, Sept 2005
58. Esterel v7 reference manual, available from www.esterel-technologies.com
59. G. Berry, Esterel on hardware. Philos. Trans. R. Soc. Lond. A **339**, 87–104 (1992)
60. S. Edwards, An Esterel compiler for large control-dominated systems. IEEE Trans. Comput. Aided Des. Integr. Circuits Syst. **21**(2), 169–183 (2002)
61. E. Closse, M. Poize, J. Pulou, P. Vernier, D. Weil, SAXO-RT: interpreting Esterel semantic on a sequential execution structure, in *Proceedings of International Workshop on Synchronous Languages, Applications, and Programming (SLAP'02)*, Electronic notes in theoretical computer science 65, Apr 2002
62. D. Potop-Butucaru, R. de Simone, Optimizations for faster execution of Esterel programs, in *Proceedings of Formal Methods and Models for Co-Design (MEMOCODE'03)*, June 2003
63. C. Passerone, C. Sansoe, L. Lavagno, P.C. McGeer, J. Martin, R. Passerone, A.L. Sangiovanni-Vincentelli, Modeling reactive systems in Java. ACM Trans. Des. Autom. Electron. Syst. **3**(4), 515–523 (1998)
64. L. Lavagno, E. Sentovich, ECL: a specification environment for system-level design, in *Proceedings of Design Automation Conference (DAC '99)*, June 1999
65. M. Antonotti, A. Ferrari, A. Flesca, A.L. Sangiovanni-Vincentelli, JESTER: an esterel-based reactive java extension for reactive embedded system development, in *Proceedings of Forum on Specification and Design Languages (FDL'00)*, Sept 2000
66. F. Boussinot, R. de Simone, The SL synchronous language. IEEE Trans. Softw. Eng. **22**(4), 256–266 (1996)
67. J.S. Young, J. MacDonald, M. Shilman, A. Tabbara, P. Hilfinger, A.R. Newton, Design and specification of embedded systems in Java using successive, formal refinement, in *Proceedings of Design Automation Conference (DAC '98)*, June1998
68. M. von der Beeck, A comparison of statecharts variants, in *Proceedings of Formal Techniques in Real Time and Fault Tolerant Systems*, LNCS 863, Sept 1994
69. F. Maraninchi, Operational and compositional semantics of synchronous automaton compositions, in *Proceedings of International Conference on Concurrency Theory (CONCUR'92)*, LNCS 630, Aug 1992
70. Open SystemC Initiative, SystemC Version 2.0 User's Guide, available at www.systemc.org

71. T. Grotker, S. Liao, G. Martin, S. Swan, *System Design with SystemC* (Kluwer, Boston/Dordrecht, 2002)
72. J. Bhasker, *A SystemC Primer* (Star Galaxy Publishing, Allentown, 2002)
73. J.G. Lee, C.M. Kyung, PrePack: predictive packetizing scheme for reducing channel traffic in transaction-level hardware/software co-emulation. IEEE Trans. Comput. Aided Des. Integr. Circuits Syst. **25**(10), 1935–1949 (2006)
74. W. Klingauf, R. Gunzel, O. Bringmann, P. Parfuntseu, M. Burton, GreenBus – a generic interconnect fabric for transaction level modelling, in *Proceedings of Design Automation Conference (DAC '06)*, July 2006
75. E. Viaud, F. Pecheux, A. Greiner, An efficient TLM/T modeling and simulation environment based on conservative parallel discrete event principles, in *Proceedings of Design, Automation and Test in Europe Conference (DATE'06)*, Mar2006
76. T. Wild, A. Herkersdorf, R. Ohlendorf, Performance evaluation for system-on-chip architectures using trace-based transaction level simulation, in *Proceedings of Design, Automation and Test in Europe Conference (DATE'06)*, Mar 2006
77. A. Habibi, S. Tahar, A. Samarah, D. Li, O. Mohamed, Efficient assertion based verification using TLM, in *Proceedings of Design, Automation and Test in Europe Conference (DATE'06)*, Mar 2006
78. G. Beltrame, D. Sciuto, C. Silvano, D. Lyonnard, C. Pilkington, Exploiting TLM and object introspection for system-level simulation, in *Proceedings of Design, Automation and Test in Europe Conference (DATE'06)*, Mar 2006
79. D. Gajski, J. Zhu, R. Domer, A. Gerstlauer, S. Zhao, *SpecC: Specification Language and Methodology* (Kluwer, Dordrecht/Boston, 2000)
80. D. Ku, G. De Micheli, HardwareC – a language for hardware design (version 2.0) CSL Technical Report CSL-TR-90-419, Stanford University, Stanford, Apr 1990
81. Handel-C Language Reference Manual, available at www.celoxica.com
82. D. Harel, H. Lachover, A. Naamad, A. Pnueli, M. Politi, R. Sherman, A. Shtull-Trauring, M. Trakhtenbrot, Statemate: a working environment for the development of complex reactive systems. IEEE Trans. Softw. Eng. **16**, 403–414 (1990)
83. J. Buck, S. Ha, E.A. Lee, D. Messerschmitt, Ptolemy: a framework for simulating and prototyping heterogeneous systems. Int. J. Comput. Simul. **4**(2), 155–182 (1994)
84. C. Hylands, E.A. Lee, J. Liu, X. Liu, S. Neuendorffer, Y. Xiong, H. Zheng, Heterogeneous concurrent modeling and design in Java, Technical Memorandum UCB/ERL M02/23, University of California, Berkeley, 2002
85. A. Girault, B. Lee, E.A. Lee, Hierarchical finite state machines with multiple concurrency models. IEEE Trans. Comput. Aided Des. Integr. Circuits Syst. **18**(6), 742–760 (1999)
86. Z. Salcic, R. Mikhael, A new method for instantaneous power system frequency measurement using reference points detection. Electr. Power Syst. Res. **55**(2), 97–102 (2000)
87. J.S. Lee, L.E. Miller, *CDMA Systems Engineering Handbook* (Artech House, Boston, 1998)
88. I. Radojevic, Z. Salcic, P. Roop, Design of heterogeneous embedded systems using DFCharts model of computation, in *Proceedings of VLSI Design*, Hyderabad, 3–7 Jan 2006
89. I. Radojevic, Z. Salcic, P. Roop, Modeling heterogeneous embedded systems in DFCharts, in *Proceedings of Forum on Design and Specification Languages* (FDL), Lausanne, 27–30 Sept 2005
90. M. Hennessy, H. Lin, Symbolic bisimulations. Theor. Comput. Sci. **138**(2), 353–389 (1995)
91. R. Alur, D. Dill, A theory of timed automata. Theor. Comput. Sci. **126**(2), 183–235 (1994)
92. S. Ramesh, Communicating reactive state machines: design, model and implementation, in *IFAC Workshop on Distributed Computer Control Systems* (Pergamon Press, Oxford, 1998)
93. S. Ramesh, Implementation of communicating reactive processes. Parallel Comput. **25**(6), 703–727 (1999)
94. G. Berry, E. Sentovich, Multiclock Esterel, in *Proceedings of Correct Hardware Design and Verification Methods (CHARME)*, LNCS 2144, Sept 2001
95. H. Patel, S. Shukla, *SystemC Kernel Extensions for Heterogeneous System Modeling: A Framework for Multi-MoC Modeling & Simulation* (Kluwer, Boston/Dordrecht, 2004)

96. H.D. Patel, S.K. Shukla, R. Bergamaschi, Heterogeneous behavioral hierarchy for system level design, in *Proceedings of Design Automation and Test in Europe*, Munich, Mar 2006
97. C. Brooks, E.A. Lee, X. Liu, S. Neuendorffer, Y. Zhao, H. Zheng (eds.), *Heterogeneous Concurrent Modeling and Design in Java (Volume 1: Introduction to Ptolemy II),* Technical Memorandum UCB/ERL M05/21, University of California, Berkeley, 15 July 2005
98. D. Atienza et al., Network-on-Chip design and synthesis outlook. Integr. VLSI J. **41**, 340–359 (2008)
99. T. Bjerregaard, S. Mahadevan, A survey of research and practices of Network-on-Chip. ACM Comput. Surv. **38**, 1–51 (2006)
100. E. Salminen, A. Kulmala, T.D. Hamalainen, Survey of network-on-chip proposals, White paper, OCP-IP, Mar 2008
101. W. Hwu, Many-core computing: can compilers and tools do the heavy lifting?, in *9th International Forum on Embedded MPSoC and Multicore*, MPSoC'09, 2009
102. Tensilica Xtensa processor, www.tensilica.com
103. Intel, Single-chip Cloud Computer Overview, Intel Corporation (2010)
104. M. Schoeberl, Schedule memory access, not threads, in *10th International Forum on Embedded MPSoC and Multicore*, MPSoC'10, 2010
105. H. Dutta et.al, Massively parallel processor architectures: a co-design approach, in *Proceedings of the 3rd International Workshop on Reconfigurable Communication Centric System-on-Chips (ReCoSoC),* Montpellier, 18–20 June 2007, pp. 61–68
106. L. Bauer et. al., KAHRISMA: a multi-grained reconfigurable multicore architecture, in *10th International Forum on Embedded MPSoC and Multicore*, MPSoC'10, 2010
107. D. Göhringer, M. Hübner, V. Schatz, J. Becker, Runtime adaptive multi-processor system-on-chip: RAMPSoC, in *IEEE International Symposium on Parallel and Distributed Processing*, 2008, pp. 1–7
108. Z. Salcic, D. Hui, P. Roop, M. Biglari-Abhari, REMIC – design of a reactive embedded microprocessor core, in *Asia-South Pacific Design Automation Conference*, Shanghai, Jan 2005
109. M.W.S. Dayaratne, P. Roop, Z. Salcic, Direct execution of Esterel using reactive microprocessors, synchronous languages, in *Applications and Programming, SLAP 05*, Edinburgh, Apr 2005
110. L.H. Yoon, P. Roop, Z. Salcic, F. Gruian, Compiling Esterel for direct execution, in *Proceedings of the Conference on Synchronous Languages, Applications and Programming, SLAP 2006*, Vienna, Mar 2006
111. Z. Salcic, D. Hui, P. Roop, M. Biglari-Abhari, HiDRA – a reactive multiprocessor architecture for heterogeneous embedded systems. Elsevier J. Microprocess. Microsyst. **30**(2), 72–85 (2006)
112. A. Malik, Z. Salcic, P. Roop, SystemJ compilation using the tandem virtual machine approach, in *ACM Transactions on Design Automation of Electronic Systems* (2009)
113. A. Malik, Z. Salcic, A. Girault, A. Walker, S.C. Lee, A customizable multiprocessor for globally asynchronous locally synchronous execution, in *Proceedings of Java Technologies for Real-time and Embedded Systems*, JTRES'09, Madrid, 2009, ACM
114. F. Gruian, P. Roop, Z. Salcic, I. Radojevic, SystemJ approach to system-level design, in *Proceedings of Methods and Models for Co-Design Conference*, Memocode 2006, Napa Valley, 2006, Piscataway, (IEEE Cat. No. 06EX1398). IEEE. 2006, pp. 149–58
115. A. Malik, Z. Salcic, P. Roop, A. Girault, SystemJ: a GALS language for system level design. Elsevier J. Comput. Lang. Syst. Struct. **36**(4), 317–344 (2010). doi:10.1016/j.cl.2010.01.001
116. Z. Salcic, P. Roop, M. Biglari-Abhari, A. Bigdeli, REFLIX: a framework of a novel processor core for reactive embedded applications. Elsevier J. Microprocess. Microsyst. **28**, 13–25 (2004)
117. X. Li, R. von Hanxleden, The Kiel Esterel Processor – a semi-custom, configurable reactive processor, in *Synchronous Programming – SYNCHRON'04*, ser. Dagstuhl Seminar Proceedings, no. 04491, ed. by S.A. Edwards, N. Halbwachs, R.v. Hanxleden, T. Stauner (Schloss Dagstuhl, Germany, 2005)

118. S. Yuan, S. Andalam, L.H. Yoong, P. Roop, Z. Salcic, STARPro – a new multithreaded direct execution platform for Esterel. EURASIP J. Embed. Syst. in press (accepted 5 Feb 2009)
119. P. Roop, Z. Salcic,S. Dayaratne, Towards direct execution of Esterel programs on reactive processors, in *Embedded Software Conference, EMSOFT'04*, Pisa, 27–29 Sept 2004
120. L. Yang, M. Biglari-Abhari, Z. Salcic, A power-efficient processor core for reactive embedded applications, in *Proceedings of Asia-South Pacific Computer Architecture Conference*, ASCAC, 2006
121. P. Petrov, A. Orailogulu, Low-power instruction bus encoding for embedded processors. IEEE Trans. Very Large Scale Integr. Syst. **12**(8), 812–826 (2004)
122. C. Zebelein, J. Falk, C. Haubelt, J. Teich ,Efficient high level modelling in the networking domain, in *Proceedings of Design, Automation and Test in Europe Conference (DATE)*, Mar 2010
123. J. Keinert, M. Streubuhr, T. Schlichter, J. Falk, J. Gladigau, C. Haubelt, J. Teich, SystemCoDesigner - an automatic ESL synthesis approach by design space exploration and behavioural synthesis for streaming applications. ACM Trans. Des. Autom. Electron. Syst. **14**, 1–23 (2009)

Index

S

T